AF564540

Agricultural Microbiology

About the Authors

Miss Monika Mishra is presently working as Sr. Research Follow in ICAR-IIPR. Her research is related to PGPRs. She has more than two years of research experience. She has isolated and characterized several potential *Trichoderma* and bacterial species against wilt pathogen of pulse crops. She has also isolated several species of many soil borne phytopathogens across the U.P. She has more than 50 NCBI database submissions of microorganism. She had been awarded gold medal during her masters. She had represent her college/university in many chess & debate competetions. She has a keen knowledge of molecular biology. She has published more than 20 research papers/reviews/ conference abstracts, popular articles & papers, book chapters.

Miss Sonika Pandey is a Sr. Research Fellow in ICAR-IIPR Kanpur. Presently she is working on the development of transgenic pigeon pea and chickpea varieties having resistance against pod borer *Helicoverpa armigera*. She has more than six years of research experience. She has worked on the secondary metabolite isolation and their profiling, nanoparticle synthesis, development of novel bioformulations with increased shelf life (*Trichoderma* spp.) and biological management of phytopathogens and insects, extraction of biodiesel and essential oils. She has one Indian patent. She has 34 research papers/review articles/ popular articles/ book chapters/books & manuals/conference abstracts/ hindi articles, magazine articles.

Dr. R.K. Mishra is presently working as Principal Scientist (Plant Pathology) is associated with Research and development activity on pulses. He involves in Host Pant Resistance (HPR), characterization and management of major biotic stresses of pigeonpea and implementation of bio-control agents based bio-intensive modules for management of biotic stresses of pulses and AICRP on pigeonpea. He identified several donors for wilt (*Fusarium udum*, race-2) resistance and Phytophthora blight in pigeonpea. He characterized potential isolates of *Trichoderma harzianum, T. asperellum, T. longibrachiatum* and developed a talc based formulations i.e. Dalhanderma (IIPRTh-31) and Dalhanderma-1 (IIPRTh-1). He handled independently project on plant pathological aspect, also conducted various training programme and organized seminars/symposia. He has more than 13 year experiences in the area of plant pathology, bio-control of plant disease management, IPM and published more than 100 research papers/reviews/ conference papers/abstracts and popular articles of international/ national repute are credited to him. Dr. Mishra has written four books, five technical bulletin, manuals and extension folders on biotic stresses management in crop plants.

Agricultural Microbiology
A Ready Reckoner

Monika Mishra
Sonika Pandey
R. K. Mishra

NEW INDIA PUBLISHING AGENCY
New Delhi – 110 034

NEW INDIA PUBLISHING AGENCY

101, Vikas Surya Plaza, CU Block, LSC Market
Pitam Pura, New Delhi – 110 034, India
Email: info@nipabooks.com
Web: www.nipabooks.com

For customer assistance, please contact
Phone: + 91-11-27 34 17 17 Fax: + 91-11- 27 34 16 16
E-Mail: feedbacks@nipabooks.com

ISBN : 978-93-89130-11-9

Composed and Designed by NIPA.

Digitally printed all across the globe.

Preface

Microbiology is the study of living organisms of microscopic size, which include bacteria, fungi, algae and protozoa and the infectious agents at the borderline of life that are called viruses. It is concerned with their form, structure, reproduction, physiology, metabolism and classification. It includes the study of their distribution in nature, their relationship to each other living organisms, their effects on human beings and on other animals and plants, their abilities to make physical and chemical changes in our environment and their reactions to physical and chemical agents. Thus, microbiology is being a part of science has given birth to the several branches *viz.* Bacteriology, Mycology, Protozoology, Virology, Parasitology, Phycology or Algology, Microbial Morphology, Microbial Physiology, Microbial Taxonomy, Microbial Genetics, Molecular Biology, Microbial Ecology, Food microbiology, Dairy microbiology, Aquatic microbiology, Industrial microbiology etc. each being pursued as a specialty in itself. Agricultural microbiology is one among them, concerned with the relationships between microbes and crops, with an emphasis on improving yields and combating plant diseases. In agricultural education, the study of microbiology has undergone tremendous changes in the past few decades in all above mentioned areas. This covers all human endeavors in the broad sense as the acquisition, transmission and absorption of knowledge for the better means and understanding of the processes which lead to the scientific farming as we all know that agriculture is a backbone of economy all over the globe. "**Agricultural Microbiology: A Ready Reckoner**" is a multiple choice question book specially designed to improve the knowledge of students and to provide them a powerful knowledge and feedback in their progress and future opportunities. Thus, primarily this book can serve as a self-assessment guide for the students who are preparing for competitive examinations specially ASRB-ARS Pre, ICAR-NET, ICAR-SRF, ICAR-JRF etc. and secondly, meant for those who appearing for UG, PG study in agriculture colleges/ agriculture institutes/ agriculture universities etc. The questions of this book are compiled from standard textbooks on History and Scope of Microbiology, Microbial Taxonomy, Microbial Ecology and Physiology, Microbial Genetics, Microbial Biotechnology, Food Microbiology, Soil Microbiology, Environmental Microbiology and Microbiological Techniques with the syllabus of ASRB-ARS Pre, ICAR-NET, ICAR-SRF, ICAR-JRF, SAU's and other competitive examinations. The main objective of this book is to help students to enrich their powerful knowledge of microbiology because it is essential for the development of students in the area of agriculture science. We hope that this book certainly will increase the knowledge of microbiology in agriculture for cracking the examinations that are essential now a day to provide complete knowledge for the proper accurate, reliable and quicker assessment. This book will also be very useful for teachers and researchers, related to microbiology subject to stimulate the success to the student's carrier. We request and welcome our esteemed Scientists, Teachers Students and Friends to send their valuable suggestions in further improvement of this book in the next issue.

Authors

Contents

Glossary

Acidiophiles - An acidophile is an organism that can or must live in an acidic environment. An acidic environment is one that has a pH below 6. Acidophiles are able to live and thrive to a highly acidic environment, particularly at pH 2.0 or below. Acidophiles are considered as an extremophile. Example sulfolobus, thiobacillus etc

Acetogenic bacteria - Group iv bacteria in methane production; they oxidized H2 by reducing CO2 to produce a acetic acid.

Allele - Alternative form of single gene; govern contrasting forms of a single character.

Anaerobes - Anaerobic may be used to describe an organism, a cell, a process or a mechanism that can function without air (i.e. air to generally mean oxygen). This is in contrast to the term aerobic, which means requiring air or free oxygen.

Aerobes - An organism able to live and reproduce only in the presence of free oxygen (***e.g***., certain bacteria and certain yeasts). Organisms that grow in the absence of free oxygen are termed anaerobes; those that grow only in the absence of oxygen are obligate, or strict, anaerobes. Some species, called facultative anaerobes, are able to grow either with or without free oxygen. Certain others, able to grow best in the presence of low amounts of oxygen, are called microaerophiles.

Amplicon - DNA sequence which act as unit of amplification

Antagonism - Antagonism refers to the action of any organism that suppress or interfere the normal growth and activity of a plant pathogen, such as the main parts of bacteria or fungi. These organisms can be used for pest control and are referred to as biological control agents.

Antibiotics - Antimicrobial agent made from microorganisms, and can kill and inhibit the growth of microorganisms, especially those that are infectious or disease-causing

Antiboidies - Special class of protein called immunoglobinig; produced in response to an antigen and specifically bind to the inducing antigen

Apoenzyme - The protein component of an enzyme that has a coenzyme cofactor

Artificial seed - A gel bead containing SE or shoot bud , necessary nutrients, etc.

Auxin - A plant growth regulator that primarily enchnace cell elongation and induces cell division in culture cells; iaa is the natural auxin.

Accesoory pigment - Accessory pigments are light-absorbing compounds, found in photosynthetic organisms, that work in conjunction with chlorophylla. They include other forms of this pigment, such as chlorophyll b in green algal and higher plant antennae, while other algae may contain chlorophyll c or d

Actinobacteria - Actinobacteria are Gram†stain†positive microorganisms with high guanine plus cytosine content in their genome (>50%). They are quite abundant in soil and aquatic sediments where they participate in the decomposition of organic matter. These bacteria are responsible for the distinctive scent of freshly exposed, moist soil.

Agar - Agar or agar-agar is a jelly-like substance, obtained from red algae.Agar is a mixture of two components: the linear polysaccharide agarose, and a heterogeneous mixture of smaller molecules called agaropectin.It forms the supporting structure in the cell walls of certain species of algae, and is released on boiling. These algae are known as agarophytes, and belong to the Rhodophyta (red algae) phylum.

Azotobacter - It is a genus of usually motile, oval or spherical bacteria that form thick-walled cysts and may produce large quantities of capsular slime. They are aerobic, free-living soil microbes which play an important role in the nitrogen cycle in nature, binding atmospheric nitrogen, Azotobacter species are Gram-negative bacteria found in neutral and alkaline soils,[1][2] in water, and in association with some plants.

Bacillus: ***Bacillus*** - Is group of rod-shaped, gram-positive, aerobic or (under some conditions) anaerobic bacteria widely found in soil and water. The term ***bacillus*** has been applied in a general sense to all cylindrical or rodlike bacteria. The largest known ***Bacillus*** species, ***B. megaterium***, is about 1.5 μm (micrometres; 1 μm = 10™6 m) across by 4 μm long. ***Bacillus*** frequently occur in chains. all Bacillus species can form dormant spores under adverse environmental conditions.

Bactericide - An agent that kills bacteria.

Bacteriochlorophyll - A modified chlorophyll that serves as the primarylight-trapping pigment in purple and green photosynthetic bacteria.

Bacteriocin - A protein produced by a bacterial strain that kills other closely related.

Bacteriophage - A virus that uses bacteria as its host; often called a phage.

Bacteriophage (phage) typing - A technique in which strains of bacteria are identified based on their susceptibility to a variety of bacteriophages

Bacteriostatic - Inhibiting the growth and reproduction of bacteria.

Bacteroid - A modified, often pleomorphic, bacterial cell within the root nodule cells of legumes; after transformation into a symbiosome it carries out nitrogen fixation.

Balanced growth - Microbial growth in which all cellular constituents are synthesized at constant rates relative to each other.

Bioterrorism - The intentional or threatened use of viruses, bacteria, fungi, or toxins from living organisms to produce death or disease in humans, animals, and plants.

Biodegradation - The breakdown of a complex chemical through biological processes that can result in minor loss of functional groups, fragmentation into larger constitutents, or complete breakdown to carbon dioxide and minerals. Often the term refers to the undesired microbial-mediated destruction of materials such as paper, paint, and textiles.

Biofilms - Organized microbial systems consisting of layers of microbial cells associated with surfaces, often with complex structural and functional characteristics. Biofilms have physical/ chemical gradients that influence microbial metabolic processes. They can form on inanimate devices.

Biogeochemical cycling - The oxidation and reduction of substances carried out by living organisms and/or abiotic processes that results in the cycling of elements within and between different parts of the ecosystem.

Bioinsecticide - A pathogen that is used to kill or disable unwanted insect pests. Bacteria, fungi, or viruses are used, either directly or after manipulation, to control insect populations.

Biologic transmission - A type of vector-borne transmission in which a pathogen goes through some morphological or physiological change within the vector.

Bioluminescence - The production of light by living cells, often through the oxidation of molecules by the enzyme luciferase.

Biomagnification - The increase in concentration of a substance in higher-level consumer organisms.

Biochemical oxygen demand (BOD) - The amount of oxygen used by organisms in water under certain standard conditions; it provides an index of the amount of microbially oxidizable organic matter present.

Carotenoids - Pigment molecules,usually yellowish in color, that are often used to aid chlorophyll in trapping light energy during photosynthesis

Catabolism - That part of metabolism in which larger, more complex molecules are broken down into smaller, simpler molecules with the release of energy.

Catabolite repression - Inhibition of the synthesis of several catabolic enzymes by a metabolite such as glucose.

Catalyst - A substance that accelerates a reaction without being permanently changed itself.

Cell cycle - The sequence of events in a cell's growth-division cycle between the end of one division and the end of the next. In eucaryotic cells, it is composed of the G1 period, the S period in which DNA and histones are synthesized, the G2period, and the M period (mitosis).

Chimera - A recombinant plasmid containing foreign DNA, which is used as a cloning vector in genetic engineering.

Chemoreceptors - Special protein receptors in the plasma membrane or periplasmic space that bind chemicals and trigger the appropriate chemotaxic response.

Chemostat - A continuous culture apparatus that feeds medium into the culture vessel at the same rate as medium containing microorganisms is removed; the medium in a chemostat contains one essential nutrient in a limiting quantity.

Chemotaxis - The pattern of microbial behavior in which the microorganism moves toward chemical attractants and/or away from repellents.

Chemotherapeutic agents - Compounds used in the treatment of disease that destroy pathogens or inhibit their growth at concentrations low enough to avoid doing undesirable damage to the host.

Chemotrophs - Organisms that obtain energy from the oxidation of chemical compounds.

Chemical oxygen demand (COD) - Theamount of chemical oxidation required to convert organic matter in water and waste water to CO_2.

Chemiosmotic hypothesis - The hypothesis that a proton gradient and an electrochemical gradient are generated by electron transport and then used to drive ATP synthesis by oxidative phosphorylation.

Chemolithotrophic autotrophs - Microorganisms that oxidize reduced inorganic compounds to derive both energy and electrons; CO_2 is their carbon source. Also called chemolithoautotrophs.

Chemoorganotrophic heterotrophs - Organisms that use organic compounds as sources of energy, hydrogen, electrons, and carbon for biosynthesis.

Chitin - A tough, resistant, nitrogencontaining polysaccharide forming the walls of certain fungi, the exoskeleton of arthropods, and the epidermal cuticle of other surface structures of certain protists and animals.

Chlamydiae - Members of the genus *Chlamydia:* gram-negative, coccoid cells that reproduce only within the cytoplasmic vesicles of host cells using a life cycle that alternates between elementary bodies and reticulate bodies.

Chlamydospore - An asexually produced, thick-walled resting spore formed by some fungi.

Chloramphenicol - A broadspectrum antibiotic that is produced by *Streptomyces venezuelae* or synthetically; it binds to the large ribosomal subunit and inhibits the peptidyl transferase reaction.

Chlorophyll - The green photosynthetic pigment that consists of a large tetrapyrrole ring with a magnesium atom in the center.

Chloroplast - A eucaryotic plastid that contains chlorophyll and is the site of photosynthesis.

Chromogen - A colorless substrate that is acted on by an enzyme to produce a colored end product.

Chromophore group - A chemical group with double bonds that absorbs visible light and gives a dye its color.

Community - An assemblage of different types of organisms or a mixture of different microbial populations.

Competent - A bacterial cell that can take up free DNA fragments and incorporate them into its genome during transformation.

Competition - An interaction between two organisms attempting to use the same resource

Colicin - A plasmid-encoded protein that is produced by enteric bacteria and binds to specific receptors on the cell envelope of sensitive target bacteria, where it may cause lysis or attack specific intracellular sites such as ribosomes.

Composting - The microbial processing of fresh organic matter under moist, aerobic conditions, resulting in the accumulation of a stable humified product, which is suitable for soil improvement and stimulation of plant growth.

Compromised host - A host with lowered resistance to infection and disease for any of several reasons. The host may be seriously debilitated (due to malnutrition, cancer, diabetes, leukemia, or another infectious disease), traumatized (from surgery or injury), immunosuppressed, or have an altered microbiota due to prolonged use of antibiotics.

Cyanobacteria - A large group of bacteria that carry out oxygenic photosynthesis using a system like that present in photosynthetic eucaryotes.

Cortex - (kor2 teks) The layer of a bacterial endospore that is thought to be particularly important in conferring heat resistance on the endospore.

Conjugants - (kon2 joo-gants) Complementary mating types that participate in a form of protozoan sexual reproduction called conjugation.

Conjugation - The form of gene transfer and recombination in bacteria that requires direct cell-to-cell contact. 2. A complex form of sexual reproduction commonly employed by protozoa.

Conjugative plasmid - A plasmid that carries the genes for sex pili and can transfer copies of itself to other bacteria during conjugation.

Cytotoxin - A toxin or antibody that has a specific toxic action upon cells; cytotoxins are named according to the cell for which they are specific.

Ceoxyribonucleic acid - The nucleic acid that constitutes the genetic material of all cellular organisms. It is a polynucleotide composed of deoxyribonucleotides connected by phosphodiester bonds.

Dipicolinic acid - A substance present at high concentrations in the bacterial endospore. It is thought to contribute to the endospore's heat resistance.

Dark-field microscopy - Microscopy in which the specimen is brightly illuminated while the background is dark.

Dark reactivation - The excision and replacement of thymine dimers in DNA that occurs in the absence of light.

Deamination - The removal ofamino groups from amino acids.

Death phase - The decrease in viable microorganisms that occurs after the completion of growth in a batch culture.

Decimal reduction time - (D or D value) The time required to kill 90% of the microorganisms or spores in a sample at a specified temperature.

Decomposer - An organism that breaks down complex materials into simpler ones, including the release of simple inorganic products. Often a decomposer such as an insect or earthworm physically reduces the size of substrate particles.

Dermatomycosis - A fungal infection of the skin; the term is a general term that comprises the various forms of tinea, and it is sometimes used to specifically refer to athlete's foot.

Desensitization - To make a sensitized or hypersensitive individual insensitive or nonreactive to a sensitizing agent.

Desert crust - A crust formed by microbial binding of sand grains in the surface zone of desert soil; crust formation primarily involves cyanobacteria.

Detergent - An organic molecule, other than a soap, that serves as a wetting agent and emulsifier; it is normally used as cleanser, but some may be used as antimicrobial agents.

Deuteromycetes - In some classification systems, the deuteromycetes or Fungi Imperfecti are a class of fungi. These organismseither lack a sexual stage or it has not yet been discovered.

Diatoms - Algal protists with siliceous cell walls called frustules. They constitute a substantial subfraction of the phytoplankton.

Diauxic growth - A biphasic growth pattern or response in which a microorganism, when exposed to two nutrients, initially uses one of them for growth and then alters its metabolism to make use of the second.

Differential media - Culture media that distinguish between groups of microorganisms based on differences in their growth and metabolic products.

Differential staining procedures - Staining procedures that divide bacteria into separate groups based on staining properties.

Disinfectant - An agent, usually chemical, that disinfects; normally, it is employed only with inanimate objects.

Disinfection - The killing, inhibition, or removal of microorganisms that may cause disease. It usually refers to the treatment of inanimate objects with chemicals.

Disinfection by-products (DBPs) - Chlorinated organic compounds formed during chlorine use for water disinfection. Many are carcinogens.

Dissimilatory nitrate reduction - The process in which some bacteria use nitrate as the electron acceptor at the end of their electron transport chain to produce ATP. The nitrate is reduced to nitrite or nitrogen gas.

Dissimilatory reduction - The use of a substance as an electron acceptor in energy generation. The acceptor (e.g., sulfate or nitrate) is reduced but not incorporated into organic matter during biosynthetic processes.

Diurnal oxygen shifts - The changes in oxygen levels that occur in waters when algae produce and use oxygen on a cyclic basis during day and night.

DNA ligase - An enzyme that joins two DNA fragments together through the formation of a new phosphodiester bond.

DNA microarrays (DNA chips) - Solid supports that have DNA attached in highly organized arrays and are normally used to evaluate gene expression.

DNA polymerase - An enzyme that synthesizes new DNA using a parental DNA strand as a template.

Dinoflagellate - An algal protist characterized by two flagella used in swimming in a spinning pattern. Many are bioluminescent and an important part of marine phytoplankton, some also are important marine pathogens.

Droplet nuclei - Small particles (1-4 m in diameter) that represent what is left from the evaporation of larger particles (10 m or more in diameter) called droplets.

Early mRNA - Messenger RNA produced early in a virus infection that codes for proteins needed to take over the host cell and manufacture viral nucleic acids.

Elementary body - A small, dormant body that serves as the agent of transmission between host cells in the chlamydial life cycle.

Elongation cycle - The cycle in protein synthesis that results in the addition of an amino acid to the growing end of a peptide chain.

Endosymbiotic theory or **hypothesis** - The theory that eucaryotic organelles such as mitochondria an chloroplasts arose when bacteria established an endosymbiotic relationship with the eukaryotic ancestor and then evolved into eucaryotic organelles.

Endotoxin - The heat-stable lipopolysaccharide in the outer membrane of the cell wall of gram-negative bacteria that is released when the bacterium lyses, or sometimes during growth, and is toxic to the host.

Enteric bacteria - Members of the family *Enterobacteriaceae* (gramnegative, peritrichous or nonmotile, facultatively anaerobic, straight rods with simple nutritional requirements); also used for bacteria that live in the intestinal tract.

Endomycorrhizal - Referring to a mutualisticassociation of fungi and plant roots in which the fungus penetrates into the root cells and arbuscules and vesicles are formed.

Entner-Doudoroff pathway - A pathway thatconverts glucose to pyruvate and glyceraldehyde 3-phosphate by producing 6-phosphogluconate and then dehydrating it.

Endospore - An extremely heat - and chemical-resistant, dormant, thick-walled spore that develops within bacteria.

Enzyme - A protein catalyst with specificity for both the reaction catalyzed and its substrates.

Electroporation - The application of an electric field to create temporary pores in the plasma membrane in order to insert DNA into the cell and transform it.

Electrophoresis - A technique that separates substances through differences in their migration rate in an electrical field due to variations in the number and kinds of charged groups they have.

Extreme barophilic bacteria - Bacteria that require a high-pressure environment to function.

Extreme environment - An environment in which physical factors such as temperature, pH, salinity, and pressure are outside of the normal range for growth of most microorganisms; these conditions allow unique organisms to survive and function.

Extremophiles - Microorganisms that grow under harsh or extreme environmental conditions such as very high temperatures or low pHs.

Extrinsic factor - An environmental factor such as temperature that influences microbial growth in food.

Evolutionary distance - A quantitative indication of the number of positions that differ between two aligned macromolecules, and presumably a measure of evolutionary similarity between molecules and organisms.

Exponential phase - The phase of the growth curve during which the microbial population is growing at a constant and maximum rate, dividing and doubling at regular intervals.

Expressed sequence tag (EST) - A partial gene sequence unique to a gene that can be used to identify and position the gene during genomic analysis.

Fimbria - A fine, hairlike protein appendage on some gram-negative bacteria that helps attach them to surfaces.

Facilitated diffusion - Diffusion across the plasma membrane that is aided by a carrier.

Facultative anaerobes - Microorganisms that do not require oxygen for growth, but do grow better in its presence.

Fermentation - An energyyielding process in which an energy substrate is oxidized without an exogenous electron acceptor. Usually organic molecules serve as both electron donors and acceptors.

Factor - The fertility factor, a plasmid that carries the genes for bacterial conjugation and makes its *E. coli* host cell the gene donor during conjugation.

Final host - The host on/in which a parasitic organism either attains sexual maturity or reproduces

Fragmentation - A type of asexual reproduction in which a thallus breaks into two or more parts, each of which forms a new thallus.

Frameshift mutations - Mutations arising from the loss or gain of a base or DNA segment, leading to a change in the codon reading frame and thus a change in the amino acids incorporated into protein.

Free energy change - The total energy change in a system that is available to do useful work as the system goes from its initial state to its final state at constant temperature and pressure.

Flavin adenine dinucleotide - An electron carrying cofactor often involved in energy production (for example, in the tricarboxylic acid cycle and the-oxidation pathway).

Fluid mosaic model - The currently accepted model of cell membranes in which the membrane is a lipid bilayer with integral proteins buried in the lipid, and peripheral proteins more loosely attached to the membrane surface.

Fluorescence microscope - A microscope that exposes a specimen to light of a specific wavelength and then forms an image from the fluorescent light produced. Usually the specimen is stained with a fluorescent dye or fluorochrome.

Fluorescent light - The light emitted by a substance when it is irradiated with light of a shorter wavelength.

Fomite - An object that is not in itself harmful but is able to harbor and transmit pathogenic organisms. Also called fomes.

Fruiting body - A specialized structure that holds sexually or asexually produced spores; found in fungi and in some bacteria.

Frustule - A silicified cell wall in the diatoms.

Fungicide - An agent that kills fungi.

Fungistatic - Inhibiting the growth and reproduction of fungi.

Fungus - Achlorophyllous, heterotrophic, spore-bearing eucaryotes with absorptive nutrition; usually, they have a walled thallus.

F value - The time in minutes at a specific temperature (usually 250°F) needed to kill a population of cells or spores.

Food-borne infection - Gastrointestinal illness caused by ingestion of microorganisms, followed by their growth within the host. Symptoms arise from tissue invasion and/or toxin production.

Food chain - The flow of energy and matter in living organisms through a producer-consumer sequence.

Food intoxication - Food poisoning caused by microbial toxins produced in a food prior to consumption. The presence of living bacteria is not required.

Food poisoning - A general term usually referring to a gastrointestinal disease caused by the ingestion of food contaminated by pathogens or their toxins.

Food web - A network of many interlinked food chains, encompassing primary producers, consumers, decomposers, and detritivores.

Gene gun - A device that uses high-pressure gas or another propellant to shoot a spray of DNA-coated microprojectiles into cells and transform them. Sometimes it is called a biolistic device.

Generalized transduction - The transfer of any part of a bacterial genome when the DNA fragment is packaged within a phage capsid by mistake.

General recombination - Recombination involving a reciprocal exchange of a pair of homologous DNA sequences; it can occur any place on the chromosome.

Generation time - The time required for a microbial population to double in number.

Genetic engineering - The deliberate modification of an organism's genetic information by directly changing its nucleic acid genome

Genome - The full set of genes present in a cell or virus; all the genetic material in an organism; a haploid set of genes in a cell.

Genomics - The study of the molecular organization of genomes, their information content, and the gene products they encode.

Genus - A well-defined group of one or more species that is clearly separate from other genera.

Geographic information system (GIS) - A data management system that organizes and displays digital map data from remote sensing and aids in the analysis of relationships between mapped features.

Germicide - An agent that kills pathogens and many nonpathogens but not necessarily bacterial endospores.

Germination - The stage following bacterial endospore activation in which the endospore breaks its dormant state. Germination is followed by outgrowth.

Gametangium (gam-¢e-tan2 je-um; pl., **gametangia**) - A structure that contains gametes or in which gametes are formed.

Gamma-proteobacteria - One of the five subgroups of proteobacteria, each with distinctive 16S rRNA sequences. This is the largest subgroup and is very diverse physiologically; many important genera are facultatively anaerobic chemoorganotrophs

Gas vacuole - A gas-filled vacuole found incyanobacteria and some other aquatic bacteria that provides flotation. It is composed of gas vesicles, which are made of protein.

Gene - A DNA segment or sequence that codes for a polypeptide, rRNA, or tRNA.

Gliding motility - A type of motility in which a microbial cell glides along when in contact with a solid surface.

Global regulatory systems - Regulatory systems that simultaneously affect many genes and pathways.

Gluconeogenesis - The synthesis of glucose from noncarbohydrate precursors such as lactate and amino acids.

Glycocalyx - A network of polysaccharides extending from the surface of bacteria and other cells.

Glycogen - A highly branchedpolysaccharide containing glucose, which is used to store carbon and energy.

Glycolysis - The anaerobic conversion of glucose to lactic acid by use of the Embden-Meyerhof pathway.

Glyoxylate cycle - A modified tricarboxylic acid cycle in which the decarboxylation reactions are bypassed by the enzymes isocitrate lyase and malate synthase; it is used to convert acetyl-CoA to succinate and other metabolites.

Gram stain - A differential staining procedure that divides bacteria into gram-positive and gramnegative groups based on their ability to retain crystal violet when decolorized with an organic solvent such as ethanol.

Grana - A stack of thylakoids in the chloroplast stroma.

Granuloma - Term applied to nodular inflammatory lesions containing phagocytic cells.

Greenhouse gases - Gases released from the earth's surface through chemical and biological processes that interact with the chemicals in the stratosphere to decrease the release of radiation from the earth. It is believed that this leads to global warming.

Griseofulvin - An antibiotic from *Penicillium griseofulvum* given orally totreat chronic dermatophytic infections of skin and nails.

Group translocation - A transport process in which a molecule is moved across a membrane by carrier proteins while being chemically altered at the same time.

Growth - An increase in cellular constituents.

Growth factors - Organic compounds that must be supplied in the diet for growth because they are essential cell components or precursors of such components and cannot be synthesized by the organism.

Halobacteria or **extreme halophiles** - A group of archaea that have an absolute dependence on high NaCl concentrations for growth and will not survive at a concentration below about 1.5 M NaCl.

Halophile - A microorganism that requires high levels of sodium chloride for growth

Heat-shock proteins - Proteins produced when cells are exposed to high temperatures or other stressful conditions. They protect the cells from damage and often aid in the proper folding of proteins.

Helical - In virology this refers to a virus with a helical capsid surrounding its nucleic acid.

Helicases - Enzymes that use ATP energy to unwind DNA ahead of the replication fork.

Heterocysts - Specialized cells produced by cyanobacteria that are the sites of nitrogen fixation.

Heteroduplex DNA - A double-stranded stretch of DNA formed by two slightly different strands that are not completely complementary.

Heterogeneous nuclear RNA (hnRNA) - The RNA transcript of DNA made by RNA polymerase II; it is then processed to form mRNA.

Heterolactic fermenters - Microorganisms that ferment sugars to form lactate, and also other products such as ethanol and CO2.

Heterotroph - An organism that uses reduced, preformed organic molecules as its principal carbon source.

Heterotrophic nitrification - Nitrification carried out by chemoheterotrophic microorganisms.

Hexon or **hexamer** - A capsomer composed of six protomers.

Hexose monophosphate pathway - See pentose phosphate pathway.

Hfr strain - A bacterial strain that donates its genes with high frequency to a recipient cell during conjugation because the F factor is integrated into the bacterial chromosome.

High-energy molecule - A molecule whose hydrolysis under standard conditions makes available a large amount of free energy (the standard free energy change is more negative than about 7 kcal/mole); a high-energy molecule readily decomposes and transfers groups such as phosphate to acceptors.

High oxygen diffusion environment - A microbial environment in close contact with air and through which oxygen can move at a rapid rate (in comparison with the slow diffusion rate of oxygen through water).

Histone - A small basic protein with large amounts of lysine and arginine that is associated with eucaryotic DNA in chromatin.

Histoplasmosis - A systemicfungal infection caused by *Histoplasma capsulatum* var *capsulatum.*

Holdfast - A structure produced by some bacteria and algae that attaches the cell to a solid object.

Holoenzyme - A complete enzyme consisting of the apoenzyme plus a cofactor.

Holozoic nutrition - In this type of nutrition, nutrients (such as bacteria) are acquired by phagocytosis and the subsequent formation of a food vacuole or phagosome.

Homolactic fermenters - Organisms that ferment sugars almost completely to lactic acid.

Horizontal gene transfer - The process in which genes are transferred from one mature, independent organism to another.

Hormogonia - Small motile fragments produced by fragmentation of filamentous cyanobacteria; used for asexual reproduction and dispersal.

Host - The body of an organism that harbors another organism. It can be viewed as a microenvironment that shelters and supports the growth and multiplication of a parasitic organism.

Host restriction - The degradation of foreign genetic material by nucleases after the genetic material enters a host cell.

Hypersensitivity - A condition of increased immune sensitivity in which the body reacts to an antigen with an exaggerated immune response that usually harms the individual. Also termed an allergy.

Hyperthermophile - A bacterium that has its growth optimum between 80°C and about 113°C. Hyperthermophiles usually do not grow well below 55°C.

Hypha - The unit of structure of most fungi and some bacteria; a tubular filament.

Icosahedral - In virology this term refers to a virus with an icosahedral capsid, which has the shape of a regular polyhedron having 20 equilateral triangular faces and 12 corners.

Identification - The process of determining that a particular isolate or organism belongs to a recognized taxon.

Idiotype - A set of one or more unique epitopes in the variable region of an immunoglobulin that distinguishes it from immunoglobulins produced by different plasma cells.

Immunofluorescence - A technique used to identify particular antigens microscopically in cells or tissues by the binding of a fluorescent antibody conjugate.

Industrial ecology - The study of the ecology of industrial societies with a major focus on material cycling, energy flow, and the ecological impacts of such societies.

Infection thread - A tubular structure formed during the infection of a root by nitrogen-fixing bacteria. The bacteria enter the root by way of the infection thread and stimulate the formation of the root nodule.

Intercalating agents - Molecules that can be inserted between the stacked bases of a DNA double helix, thereby distorting the DNA and inducing insertion and deletion mutations.

Intranuclear inclusion body - A structure found within cells infected with the cytomegalovirus.

Intrinsic factors - Food-related factors such asmoisture, pH, and available nutrients that influence microbial growth.

Intron - A noncoding intervening sequence in a split or interrupted gene, which codes for RNA that is missing from the final RNA product.

Ionizing radiation - Radiation of very short wavelength or high energy that causes atoms to lose electrons or ionize.

Invasiveness - The ability of a microorganism to enter a host, grow and reproduce within the host, and spread throughout its body.

Integration - The incorporation of one DNA segment into a second DNA molecule to form a new hybrid DNA. Integration occurs during such processes as genetic recombination, episome incorporation into host DNA, and prophage insertion into the bacterial chromosome.

Jaccard coefficient (*SJ*) - An association coefficient used in numerical taxonomy; it is the proportion of characters that match, excluding those that both organisms lack.

Kinetoplast - A special structure in the mitochondrion of kinetoplastid protozoa. It contains the mitochondrial DNA.

Kirby-Bauer method - A disk diffusion test to determine the susceptibility of a microorganism to chemotherapeutic agents.

Koch's postulates - A set of rules for proving that a microorganism causes a particular disease.

Lactic acid fermentation - A fermentation that produces lactic acid as the sole or primary product.

Lager - Pertaining to the process of aging beers to allow flavor development.

Lag phase - A period following the introduction of microorganisms into fresh culture medium when there is no increase in cell numbers or mass during batch culture.

Laminarin - One of the principal storage products of the golden-brown algae; a polymer of glucose.

Lancefield system (group; lans2 feld) - One of the serologically distinguishable groups (as group A, group B) into which streptococci can be divided.

Land farming - The addition of waste material,such as a hydrocarbon waste, to the soil surface so that it will be degraded. The soil may be moistened or mixed to stimulate the desired degradation process.

Langerhans cell - Cell found in the skin that internalizes antigen and moves in the lymph to lymph nodes where it differentiates into a dendritic cell.

Late mRNA - Messenger RNA produced later in a virus infection, which codes for proteins needed in capsid construction and virus release.

Latent period - The initial phase in the one-step growth experiment in which no phages are released.

Latent virus infections - Virus infections in which the virus quits reproducing and remains dormant for a period before becoming active again.

Leader sequence - A nontranslated sequence at the 52 end of mRNA that lies between the operator and the initiation codon; it aids in the initiation and regulation of transcription.

Lectin complement pathway - The lectin pathway for complement activation is triggered by the binding of a serum lectin (mannan-binding lectin; MBL) to mannose-containing proteins or to carbohydrates on viruses or bacteria.

Lethal dose 50 (LD_{50}) - Refers to the dose ornumber of organisms that will kill 50% of an experimental group of hosts within a specified time period.

Leukocyte - Any colorless white blood cell. Can be classified into granular and agranular lymphocytes.

Lichen - An organism composed of a fungus and either green algae or cyanobacteria in a symbiotic association.

Liebig's law of the minimum - Living organisms and populations will grow until lack of a resource begins to limit further growth.

Lipopolysaccharide - A molecule containing both lipid and polysaccharide, which is important in the outer membrane of the gram-negative cell wall.

Lithotroph - An organism that uses reduced inorganic compounds as its electron source.

Low oxygen diffusion environment - An aquatic environment in which microorganisms are surrounded by deep water layers that limit oxygen diffusion to the cell surface. In contrast, microorganisms in thin water films have good oxygen transfer from air to the cell surface.

LPS-binding protein - A special plasma protein that binds bacterial lipopolysaccharides and then attaches to receptors on monocytes, macrophages, and other cells. This triggers the release of IL-1 and other cytokines that stimulate the development of fever and additional endotoxin effects.

Lysis - The rupture or physical disintegration of a cell.

Lysogenic - See lysogens.

Lysogens - Bacteria that are carrying a viral prophage and have the potential of producing bacteriophages under the proper conditions.

Lysogeny - The state in which a phage genome remains within the bacterial host cell after infection and reproduces along with it rather than taking control of the host and destroying it.

Lysosome - A spherical membranous eucaryotic organelle that contains hydrolytic enzymes and is responsible for the intracellular digestion of substances.

Lysozyme - An enzyme that degrades peptidoglycan by hydrolyzing the (1-4) bond that joins *N*-acetylmuramic acid and Nacetyl glucosamine.

Lytic cycle - A virus life cycle that results in the lysis of the host cell.

M cell - Specialized cell of the intestinal mucosa and other sites, such as the urogenital tract, that delivers the antigen from the apical face of the cell to lymphocytes clustered within the pocket in its basolateral face.

Macrolide antibiotic - An antibiotic containing a macrolide ring, a large lactone ring with multiple keto and hydroxyl groups, linked to one or more sugars.

Macromolecule - A large molecule that is a polymer of smaller units joined together.

Macromolecule vaccine - A vaccine made of specific, purified macromolecules derived from pathogenic microorganisms.

Macronucleus - The larger of the two nuclei in ciliate protozoa. It is normally polyploid and directs the routine activities of the cell.

Macrophage - The name for a large mononuclear phagocytic cell, present in blood, lymph, and other tissues. Macrophages are derived from monocytes. They phagocytose and destroy pathogens; some macrophages also activate B cells and T cells.

Maintenance energy - The energy a cell requires simply to maintain itself or remain alive and functioning properly. It does not include the energy needed for either growth or reproduction.

Maduromycosis - A subcutaneous fungal infection caused by *Madurella mycetoma;* also termed an eumycotic mycetoma.

Madurose - The sugar derivative 3-O-methyl-Dgalactose, which is characteristic of several actinomycete genera that are collectively called maduromycetes.

Magnetosomes - Magnetite particles in magnetotactic bacteria that are tiny magnets and allow the bacteria to orient themselves in magnetic fields.

Malt - Grain soaked in water to soften it, induce germination, and activate its enzymes. The malt is then used in brewing and distilling.

Mash - The soluble materials released from germinated grains and prepared as a microbial growth medium.

Mesophile - A microorganism with a growth optimum around 20 to 45°C, a minimum of 15 to 20°C, and a maximum about 45°C or lower.

Messenger RNA (mRNA) - Single-stranded RNA synthesized from a DNA template during transcription that binds to ribosomes and directs the synthesis of protein.

Metabolic channeling - The localization of metabolites and enzymes in different parts of a cell.

Metabolic control engineering - Modification of the controls for biosynthetic pathways without altering the pathways themselves in order to improve process efficiency.

Metabolic pathway engineering (MPE) - The use of molecular techniques to improve the efficiency of pathways that synthesize specific products.

Metabolism - The total of all chemical reactions in the cell; almost all are enzyme catalyzed.

Metachromatic granules - Granules of polyphosphate in the cytoplasm of some bacteria that appear a different color when stained with a blue basic dye. They are storage reservoirs for phosphate. Sometimes called volutin granules.

Metastasis - The transfer of a disease like cancer from one organ to another not directly connected with it.

Methanogens - Strictly anaerobic archaeons that derive energy by converting CO_2, H2, formate, acetate, and other compounds to either methane or methane and CO_2.

Methylotroph - A bacterium that uses reduced one-carbon compounds such as methane and methanol as its sole source of carbon and energy.

Michaelis constant - A kinetic constant for an enzyme reaction that equals the substrate concentration required for the enzyme to operate at half maximal velocity.

Microaerophile - A microorganism that requires low levels of oxygen for growth, around 2-10%, but is damaged by normal atmospheric oxygen levels.

Microarray technology - Profiling of gene expression by measuring binding of RNA from growing cells to an array of functionspecific oligonucleotides attached to an inert surface.

Microbial dietary adjuvant - A substance added to the diet to stimulate specific microbial processes and populations.

Microbial ecology - The study of microorganisms in their natural environments, with a major emphasis on physical conditions, processes, and interactions that occur on the scale of individual microbial cells.

Microbial transformation - *See* bioconversion.

Microbial loop - The mineralization of organic matter synthesized by photosynthetic phytoplankton through the activity of microorganisms such as bacteria and protozoa. This process "loops" minerals and carbon dioxide back for reuse by the primary producers and makes the organic matter unavailable to higher consumers.

Microbial mat - A firm structure of layered microorganisms with complementary physiological activities that can develop on surfaces in aquatic environments.

Microbiology - The study of organisms that are usually too small to be seen with the naked eye. Special techniques are required to isolate and grow them.

Microbivory - The use of microorganisms as a food source by organisms that can ingest or phagocytose them.

Microenvironment - The immediate environment surrounding a microbial cell or other structure, such as a root.

Microfilaments - Protein filaments, about 4 to 7 nm in diameter, that are present in the cytoplasmic matrix of eukaryotic cells and play a role in cell structure and motion.

Micronucleus - The smaller of the two nuclei in ciliate protozoa. Micronuclei are diploid and involved only in genetic recombination and the regeneration of macronuclei.

Micronutrients - Nutrients such as zinc, manganese, and copper that are required in very small quantities for growth and reproduction. Also called trace elements. organism that is too small to be seen clearly with the naked eye.

Microtubules - Small cylinders, about 25 nm in diameter, made of tubulin proteins and present in the cytoplasmic matrix and flagella of eucaryotic cells; they are involved in cell structure and movement.

Mycologist - A person specializing in mycology; a student of mycology.

Mycology - The science and study of fungi.

Mycoplasma - Bacteria that are members of the class *Mollicutes* and order *Mycoplasmatales;* they lack cell walls and cannot synthesize peptidoglycan precursors; most require sterols for growth; they are the smallest organisms capable of independent reproduction.

Mycorrhizosphere - The region around a mycorrhizal fungus in which nutrients released from the fungus increase the microbial population and its activities.

Mycosis - Any disease caused by a fungus.

Mycotoxicology - The study of fungal toxins and their effects on various organisms.

Myxobacteria - A group of gram-negative, aerobic soil bacteria characterized by gliding motility, a complex life cycle with the production of fruiting bodies, and the formation of myxospores.

Myxospores - Special dormant spores formed by the myxobacteria.

Must - The juices of fruits, including grapes, that can be fermented for the production of alcohol.

Mutagen - A chemical or physical agent that causes mutations.

Mutation - A permanent, heritable change in the genetic material.

Mutualism - A type of symbiosis in which both partners gain from the association and are unable to survive without it. The mutualist and the host are metabolically dependent on each other.

Mutualist - An organism associated with another in a relationship that is beneficial to both (and often obligatory).

Mycelium - A mass of branching hyphae found in fungi and some bacteria.

Narrow-spectrum drugs - Chemotherapeutic agents that are effective only against a limited variety of microorganisms.

Natural attenuation - The decrease in the level of an enviromental contaminant that results from natural chemical, physical, and biological processes.

Natural classification - A classification system that arranges organisms into groups whose members share many characteristics and reflect as much as possible the biological nature of organisms.

Nocardioforms - Bacteria that resemble members of the genus *Nocardia;* they develop a substrate mycelium that readily breaks up into rods and coccoid elements (a quality sometimes called fugacity).

Nomenclature - The branch of taxonomy concerned with the assignment of names to taxonomic groups in agreement with published rules.

Noncyclic photophosphorylation - The process in which light energy is used to make ATP when electrons are moved from water to NADP during photosynthesis; both photosystem I and photosystem II are involved.

Nucleus - The eucaryotic organelle enclosed by a double-membrane envelope that contains the cell's chromosomes.

Numerical aperture - The property of a microscope lens that determines how much light can enter and how great a resolution the lens can provide.

Numerical taxonomy - The grouping by numerical methods of taxonomic units into taxa based on their character states.

Nutrient - A substance that supports growth and reproduction.

Nystatin - A polyene antibiotic from *Streptomyces noursei* that is used in the treatment of *Candida* infections of the skin, vagina, and alimentary tract.

Negative staining - A staining procedure in which a dye is used to make the background dark while the specimen is unstained.

Negri bodies - Masses of viruses or unassembled viral subunits found within the brain neurons of rabies-infected animals.

Neustonic - The microorganisms that live at the atmospheric interface of a water body.

Neutrophil - A mature white blood cell in the granulocyte lineage formed in bone marrow.It has a nucleus with three to five lobes and is very phagocytic.

Neutrophile - Microorganisms that grow best at a neutral pH range between pH 5.5 and 8.0.

Niche (nich) - The function of an organism in a complex system, including place of the organism, the resources used in a given location, and the time of use.

Nicotinamide adenine dinucleotide - An electron-carrying coenzyme; it is particularly important in catabolic processes and usually donates its electrons to the electron transport chain under aerobic conditions.

Nicotinamide adenine dinucleotide phosphate - An electron-carrying coenzyme that most often participates as an electron carrier in biosynthetic metabolism.

Nitrification - The oxidation of ammonia to nitrate.

Nitrifying bacteria - Chemolithotrophic, gram-negative bacteria that are members of the family *Nitrobacteriaceae* and convert ammonia to nitrate and nitrite to nitrate.

Nitrogenase - The enzyme that catalyzes biological nitrogen fixation

Nitrogen fixation - The metabolic process in which atmospheric molecular nitrogen is reduced to ammonia; carried out by cyanobacteria, *Rhizobium,* and other nitrogen-fixing bacteria.

Nitrogen oxygen demand (*NOD*) - The demand for oxygen in sewage treatment, caused by nitrifying microorganisms.

Nitrogen saturation point - The point at which mineral nitrogen, when added to an ecosystem, can no longer be incorporated into organic matter through biological processes.

Nocardioforms - Bacteria that resemble members of the genus *Nocardia;* they develop a substrate mycelium that readily breaks up into rods and coccoid elements (a quality sometimes called fugacity).

Nomenclature - The branch of taxonomy concerned with the assignment of names to taxonomic groups in agreement with published rules.

Noncyclic photophosphorylation - The process in which light energy is used to make ATP when electrons are moved from water to NADP_during photosynthesis; both process of forming a hybrid double-stranded DNA molecule using a heated mixture of singlestranded DNAs from two different sources; if the sequences are fairly complementary, stable hybrids will form.

Nucleocapsid - The nucleic acid and its surrounding protein coat or capsid; the basic unit of virion structure.

Nucleoid - An irregularly shaped region in the procaryotic cell that contains its genetic material.

Nucleolus - The organelle, located within the eucaryotic nucleus and not bounded by a membrane, that is the location of ribosomal RNA synthesis and the assembly of ribosomal subunits.

Nucleoside - A combination of ribose or deoxyribose with a purine or pyrimidine base.

Nucleosome - A complex of histones and DNA found in eucaryotic chromatin; the DNA is wrapped around the surface of the beadlike histone complex.

Obligate aerobes - Organisms that grow only in the presence of oxygen.

Obligate anaerobes - Microorganisms that cannot tolerate the presence of oxygen and die when exposed to it.

Odontopathogens - Dental pathogens.

Okazaki fragments - Short stretches of polynucleotides produced during discontinuous DNA replication.

Oligotrophic environment - An environment containing low levels of nutrients, particularly nutrients that support microbial growth.

One-step growth experiment - An experiment used to study the reproduction of lytic phages in which one round of phage reproduction occurs and ends with the lysis of the host bacterial population.

Onychomycosis - A fungal infection of the nail plate producing nails that are opaque, white, thickened, friable, and brittle. Also called ringworm of the nails and tinea unguium. Caused by *Trichophyton* and other fungi such as *C. albicans.*

Oocyst - Cyst formed around a zygote of malaria and related protozoa.

Oogonia - Mitotically dividing female structures that produce primary oocytes and gametes.

Oomycetes - A collective name for members of the division *Oomycota;* also known as the water molds.

Open reading frame (ORF) - A reading frame sequence not interrupted by a stop codon; it is usually determined by nucleic acid sequencing studies.

Operator - The segment of DNA to which the repressor protein binds; it controls the expression of the genes adjacent to it.

Operon - The sequence of bases in DNA that contains one or more structural genes together with the operator controlling their expression.

Opportunistic microorganism or **pathogen** - A microorganism that is usually free-living or a part of the host's normal microbiota, but which may become pathogenic under certain circumstances, such as when the immune system is compromised.

Opsonization - The action of opsonins in making bacteria and other cells more readily phagocytosed. Antibodies, complement (especially C3b), and fibronectin are potent opsonins.

Organelle - A structure within or on a cell that performs specific functions and is related to the cell in a way similar to that of an organ to the body.

Organotrophs - Organisms that use reduced organic compounds as their electron source.

Osmophilic microorganisms - Microorganisms that grow best in or on media of high solute concentration.

Osmosis - The movement of water across a selectively permeable membrane from a dilute solution (higher water concentration) to a more concentrated solution.

Osmotolerant - Organisms that grow over a fairly wide range of water activity or solute concentration.

Outbreak - The sudden, unexpected occurrence of a disease in a given population.

Outer membrane - A special membrane located outside the peptidoglycan layer in the cell walls of gram-negative bacteria.

Oxidation-reduction (redox) reactions - Reactions involving electron transfers; the reductant donates electrons to an oxidant.

Oxidative phosphorylation - The synthesis of ATP from ADP using energy made available during electron transport.

Oxidizing agent or oxidant - The electron acceptor in an oxidation-reduction reaction.

Oxygenic photosynthesis - Photosynthesis that oxidizes water to form oxygen; the form of photosynthesis characteristic of eucaryotic algae and cyanobacteria.

Parasite - An organism that lives on or within another organism (the host) and benefits from the association while harming its host. Often the parasite obtains nutrients from the host.

Parasitism - A type of symbiosis in which one organism adversely affects the other (the host), but cannot live without it.

Pellicle - A relatively rigid layer of proteinaceous elements just beneath the plasma membrane in many protozoa and algae. The plasma membrane is sometimes considered part of the pellicle

Pentose phosphate pathway - The pathway that oxidizes glucose 6-phosphate to ribulose 5-phosphate and then converts it to a variety of three to seven carbon sugars; it forms several important products (NADPH for biosynthesis, pentoses, and other sugars) and also can be used to degrade glucose to CO_2.

Parasitism - A type of symbiosis in which one organism adversely affects the other (the host), but cannot live without it.

Peptide interbridgev - A short peptide chain that connects the tetrapeptide chains in some peptidoglycans.

Peptidoglycan - A large polymer composed of long chains of alternating *N*-acetylglucosamine and *N*-acetylmuramic acid residues. The polysaccharide chains are linked to each other through connections between tetrapeptide chains attached to the *N*acetylmuramic acids. It provides much of the strength and rigidity possessed by bacterial cell walls.

Peptidyl transferase - The enzyme that catalyzes the transpeptidation reaction in protein synthesis; in this reaction, an amino acid is added to the growing peptide chain.

Peptones - Water-soluble digests or hydrolysates of proteins that are used in the preparation of culture media.

Pasteur effect - The decrease in the rate of sugar catabolism and change to aerobic respiration that occurs when microorganisms are switched from anaerobic to aerobic conditions.

Pasteurization - The process of heating milk and other liquids to destroy microorganisms that can cause spoilage or disease.

Pathogen - Any virus, bacterium, or other agent that causes disease.

Pathogenicity - The condition or quality of being pathogenic, or the ability to cause disease.

Period of infectivity - Refers to the time during which the source of an infectious disease is infectious or is disseminating the pathogen. branches, much like a tree in shape, that shows phylogenetic relationships between groups of organisms and sometimes also indicates the evolutionary development of groups.

Phytoplankton - A community of floating photosynthetic organisms, largely composed of algae and cyanobacteria.

Phytoremediation - The use of plants and their associated microorganisms to remove, contain, or degrade environmental contaminants.

Piedra - A fungal disease of the hair in which white or black nodules of fungi form on the shafts.

Pinocytosis - The endocytotic process in which a cell encloses a small amount of the surrounding liquid and its solutes in tiny pinocytotic vesicles or pinosomes.

Pitched - Pertaining to inoculation of a nutrient medium with yeast, for example, in beer brewing.

Plague - An acute febrile, infectious disease,caused by the bacillus *Yersinia pestis,* which has a high mortality rate; the two major types are bubonic plague and pneumonic plague.

Plankton - Free-floating, mostly microscopic microorganisms that can be found in almost all waters; a collective name.

Photolithotrophic autotrophs - Organisms that use light energy, an inorganic electron source (e.g., H_2O, H_2, H_2S), and CO_2 as a carbon source.

Photoorganotrophic heterotrophs - Microorganisms that use light energy and organic electron donors, andalso employ simple organic molecules rather than CO_2 as their carbon source.

Photoreactivation - The process in which blue light is used by a photoreactivating enzyme to repair thymine dimmers in DNA by splitting them apart.

Photosynthesis - The trapping of light energy and its conversion to chemical energy, which is then used to reduce CO_2 and incorporate it into organic form.

Photosystem I - The photosystem in eukaryotic cells that absorbs longer wavelength light, usually greater than about 680 nm, and transfers the energy to chlorophyll P700 during photosynthesis; it is involved in both cyclic photophosphorylation and noncyclic photophosphorylation.

Photosystem II - The photosystem in eucaryotic cells that absorbs shorter wavelength light, usually less than 680 nm, and transfers the energy to chlorophyll P680 during photosynthesis; it participates in noncyclic photophosphorylation

Phycobiliproteins - Photosynthetic pigments that are composed of proteins with attached tetrapyrroles; they are often found in cyanobacteria and red algae.

Phycobilisomes - Special particles on the membranes of cyanobacteria that contain photosynthetic pigments and electron transport chains.

Phycobiont - The algal or cyanobacterial partner in a lichen.

Phycocyanin - A blue phycobiliprotein pigment used to trap light energy during photosynthesis.

Phycoerythrin - A red photosynthetic phycobiliprotein pigment used to trap light energy.

Phycology - The study of algae; algology.

Phyllosphere - The surface of plant leaves.

Phylogenetic or **phyletic classification system** - A classification system based on evolutionary relationships rather than the general similarity of contemporary characteristics.

Plasmid fingerprinting - A technique used to identify microbial isolates as belonging to the same strain because they contain the same number of plasmids with the identical molecular weights and similar phenotypes.

Plasmodial (acellular) slime mold - A member of the division *Myxomycota* that exists as a thin, streaming, multinucleate mass of protoplasm, which creeps along in an amoeboid fashion.

Plasmodium - A stage in the life cycle of myxomycetes (plasmodial slime molds); a multinucleate mass of protoplasm surrounded by a membrane. Also, a parasite of the genus *Plasmodium*

Plasmolysis - The process in which water osmotically leaves a cell, which causes the cytoplasm to shrivel up and pull the plasma membrane away from the cell wall.

Plastid - A cytoplasmic organelle of algae and higher plants that contains pigments such as chlorophyll, stores food reserves, and often carries out processes such as photosynthesis.

Pleomorphic - Refers to bacteria that are variable in shape and lack a single, characteristic form.

Plus strand or **positive strand** - The virus nucleicacid strand that is equivalent in base sequence to the viral mRNA.

Q fever - An acute zoonotic disease caused by the rickettsia *Coxiella burnetii.*

Quellung reaction - The increase in visibility or the swelling of the capsule of a microorganism in the presence of antibodies against capsular antigens.

Quorum sensing - The process in which bacteria monitor their own population density by sensing the levels of signal molecules that are released by the microorganisms. When these signal molecules reach a threshold concentration, the population density has attained a critical level or quorum, and quorumdependent genes are expressed.

Racking - The removal of sediments from wine bottles.

Radappertization - The use of gamma rays from a cobalt source for control of microorganisms in foods.

Reading frame - The way in which nucleotides in DNA and mRNA are grouped into codons or groups of three for reading the message contained in the nucleotide sequence.

Recombinant DNA technology - The techniques used in carrying out genetic engineering; they involve the identification and isolation of a specific gene, the insertion of the gene into a vector such as a plasmid to form a recombinant molecule, and the production of large quantities of the gene and its products

Recombination repair - A DNA repair process that repairs damaged DNA when there is no remaining template; a piece of DNA from a sister molecule is used.

Repressible enzyme - An enzyme whose level drops in the presence of a small molecule, usually an end product of its metabolic pathway.

Repressible enzyme - An enzyme whose level drops in the presence of a small molecule, usually an end product of its metabolic pathway.

Residuesphere - The region surrounding organic matter such as a seed or plant part in which microbial growth is stimulated by increased organic matter availability.

Resolution - The ability of a microscope to separate or distinguish between small objects that are close together.

Respiration - An energy-yielding process in which the energy substrate is oxidized using an exogenous or externally derived electron acceptor.

Rhizosphere - A region around the plant root where materials released from the root increase the microbial population and its activities.

Rho factor (ro) - The protein that helps RNA polymerase dissociate from the terminator after it has stopped transcription.

Rhoptry - Saclike, electron dense structure in the anterior portion of a zoite of a member of the phylum *Apicomplexa;* perhaps involved in the penetration of host cells.

Ribonucleic acid (RNA) - A polynucleotide composed of ribonucleotides joined by phosphodiester bridges.

Ribosomal RNA (rRNA) - The RNA present in ribosomes; ribosomes contain several sizes of single-stranded rRNA that contribute to ribosome structure and are also directly involved in the mechanism of protein synthesis.

Ribosome - The organelle where protein synthesis occurs; the message encoded in mRNA is translated here.

Ribulose-1,5-bisphosphate carboxylase - The enzyme that catalyzes the incorporation of CO_2 in the Calvin cycle.

Rolling-circle mechanism - A mode of DNA replication in which the replication fork moves around a circular DNA molecule, displacing a strand to give a tail that is also copied to produce a new double-stranded DNA.

Root nodule - Gall-like structures on roots that contain endosymbiotic nitrogen-fixing bacteria (e.g., *Rhizobium* or *Bradyrhizobium* is present in legume nodules).

Sanitization - Reduction of the microbial population on an inanimate object to levels judged safe by public health standards; usually, the object is cleaned.

Saprophyte - An organism that takes up nonliving organic nutrients in dissolved form and usually grows on decomposing organic matter.

Saprozoic nutrition - Having the type of nutrition in which organic nutrients are taken up in dissolved form; normally refers to animals or animal-like organisms.

Sanitization - Reduction of the microbial population on an inanimate object to levels judged safe by public health standards; usually, the object is cleaned.

Saprophyte - An organism that takes up nonliving organic nutrients in dissolved form and usually grows on decomposing organic matter.

Saprozoic nutrition - Having the type of nutrition in which organic nutrients are taken up in dissolved form; normally refers to animals or animal-like organisms.

Secondary metabolites - Products of metabolism that are synthesized after growth has been completed.

Secondary treatment - The biological degradation of dissolved organic matter in the process of sewage treatment; the organic material is either mineralized or changed to settleable solids.

Second law of thermodynamics - Physical and chemical processes proceed in such a way that the entropy of the universe (the system and its surroundings) increases to the maximum possible.

Secretory vacuole - In protists and some animals, these organelles usually contain specific enzymes that perform various functions such as excystation. Their contents are released to the cell exterior during exocytosis.

Segmented genome - A virus genome that is divided into several parts or fragments, each probably coding for the synthesis of a single polypeptide; segmented genomes are very common among the RNA viruses.

Selective media - Culture media that favor the growth of specific microorganisms; this may be accomplished by inhibiting the growth of undesired microorganisms.

Selective toxicity - The ability of a chemotherapeutic agent to kill or inhibit a microbial pathogen while damaging the host as little as possible.

Sorus - A type of fruiting structure composed of a mass of spores or sporangia. *565*

SOS repair - A complex, inducible repair process that is used to repair DNA when extensive damage has occurred.

Source - The location or object from which a pathogen is immediately transmitted to the host, either directly or through an intermediate agent.

Southern blotting technique - The procedure used to isolate and identify DNA fragments from a complex mixture. The isolated, denatured fragments are transferred from an agarose electrophoretic gel to a nitrocellulose filter and identified by hybridization with probes

Sorocarp - The fruiting structure of the Acrasiomycetes.

Sludge - A general term for the precipitated solid matter produced during water and sewage treatment; solid particles composed of organic matter and microorganisms that are involved in aerobic sewage treatment (activated sludge).

Slime - The viscous extracellular glycoproteins or glycolipids produced by staphylococci and *Pseudomonas aeruginosa* bacteria that allows them to adhere to smooth surfaces such as prosthetic medical devices and catheters. More generally, the term often refers to an easily removed, diffuse, unorganized layer of extracellular material that surrounds a bacterial cell.

Slime layer - A layer of diffuse, unorganized, easily removed material lying outside the bacterial cell wall.

Slime mold - A common term for members of the divisions *Acrasiomycota* and *Myxomycota.*

Slow sand filter - A bed of sand through which water slowly flows; the gelatinous microbial layer on the sand grain surface removes waterborne microorganisms, particularly *Giardia,* by adhesion to the gel. This type of filter is used in some water purification plants.

Slow virus disease A progressive, pathological process caused by a transmissible agent (virus or prion) that remains clinically silent during a prolonged incubation period of months to years after which

Slash-and-burn agriculture - The cutting down and burning of tropical vegetation to make mineral nutrients available for use by introduced agricultural crops.

S-layer - A regularly structured layer composed of protein or glycoprotein that lies on the surface of many bacteria. It may protect the bacterium and help give it shape and rigidity.

Settling basin - A basin used during water purification to chemically precipitate out fine particles, microorganisms, and organic material by coagulation or flocculation.

Sex pilus - A thin protein appendage required for bacterial mating or conjugation. The cell with sex pili donates DNA to recipient cells.

Superinfection - A new bacterial or fungal infection of a patient that is resistant to the drug(s) being used for treatment.

Superoxide dismutase (SOD) - An enzyme that protects many microorganisms by catalyzing the destruction of the toxic superoxide radical.

Sulfonamide - A chemotherapeutic agent that has the SO2-NH2 group and is a derivative of sulfanilamide.

Substrate-level phosphorylation - The synthesis of ATP from ADP by phosphorylation coupled with the exergonic breakdown of a high-energy organic substrate molecule.

Subsurface biosphere - The region below the plant root zone where microbial populations can grow and function.

Sulfate reduction - The process of sulfate use as an oxidizing agent, which results in the accumulation of reduced forms of sulfur such as sulfide, or incorporation of sulfur into organic molecules, usually as sulfhydryl groups.

Stromatolite - Dome-like microbial mat communities consisting of filamentous photosynthetic bacteria and occluded sediments (often calcareous or siliceous). They usually have a laminar structure. Many are fossilized, but some modern forms occur.

Stroma - The chloroplast matrix that is the location of the photosynthetic carbon dioxide fixation reactions.

Symbiosis - The living together or close association of two dissimilar organisms, each of these organisms being known as a symbiont.

Symbiosome - The final nitrogen-fixing form of *Rhizobium* that is active within root nodule cells.

Syntrophism - The association in which the growth of one organism either depends on, or is improved by, the provision of one or more growth factors or nutrients by a neighboring organism. Sometimes both organisms benefit. This type of mutualism is also known as cross-feeding or the satellite phenomenon.

Taxon - A group into which related organisms are classified.

Taxonomy - The science of biological classification; it consists of three parts: classification, nomenclature, and identification.

T cell or **T lymphocyte** - A type of lymphocyte derived from bone marrow stem cells that matures into an immunologically competent cell under the influence of the thymus. T cells are involved in a variety of cell-mediated immune reactions.

Thermoacidophiles - A group of bacteria that grow best at acid pHs and high temperatures; they are members of the *Archaea.*

Theory - A set of principles and concepts that have survived rigorous testing and that provide a systematic account of some aspect of nature.

Thermal death time (TDT) - The shortest period of time needed to kill all the organisms in a microbial population at a specified temperature and under defined conditions.

Thylakoid - A flattened sac in the chloroplast stroma that contains photosynthetic pigments and the photosynthetic electron transport chain; light energy is trapped and used to form ATP and NAD(P)H in the thylakoid membrane.

Thymine - The pyrimidine 5-methyluracil that is found in nucleosides, nucleotides, and DNA.

Thermophile - A microorganism that can grow at temperatures of 55°C or higher; the minimum is usually around 45°C.

Teichoic acids - Polymers of glycerol or ribitol joined by phosphates; they are found in the cell walls of gram-positive bacteria.

Temperate phages - Bacteriophages that can infect bacteria and establish a lysogenic relationship rather than immediately lysing their hosts.

Template strand - A strand of DNA or RNA that specifies the base sequence of a newly synthesized complementary strand of DNA or RNA.

Terminator - A sequence that marks the end of a gene and stops transcription.

Tertiary treatment - The removal from sewage of inorganic nutrients, heavy metals, viruses, etc., by chemical and biological means after microorganisms have degraded dissolved organic material during secondary sewage treatment.

Transposon - A DNA segment that carries the genes required for transposition and moves about the chromosome; if it contains genes other than those required for transposition, it may be called a composite transposon. Often the name is reserved only for transposable elements that also contain genes unrelated to transposition.

Transversion mutations - Mutations that result from the substitution of a purine base for the normal pyrimidine or a pyrimidine for the normal purine.

Transamination - The removal of amino acid's amino group by transferring it to an-keto acid acceptor.

Transcriptase - An enzyme that catalyzes transcription; in viruses with RNA genomes, this enzyme is an RNA-dependent RNA polymerase that is used to make RNA copies of the RNA genomes.

Transcription - The process in which single-stranded RNA with a base sequence complementary to the template strand of DNA or RNA is synthesized.

Transduction - The transfer of genes between bacteria by bacteriophages.

Transfer host - A host that is not necessary for the completion of a parasite's life cycle, but is used as a vehicle for reaching a final host.

Transfer RNA (tRNA) - A small RNA that binds an amino acid and delivers it to the ribosome for incorporation into a polypeptide chain during protein synthesis.

Transformation - A mode of gene transfer in bacteria in which a piece of free DNA is taken up by a bacterial cell and integrated into the recipient genome.

Transgenic animal or **plant** - An animal or plant that has gained new genetic information from the insertion of foreign DNA. It may be produced by such techniques as injecting DNA into animal eggs, electroporation of mammalian cells and plant cell protoplasts, or shooting DNA into plant cells with a gene gun.

Transition mutations - Mutations that involve the substitution of a different purine base for the purine present at the site of the mutation or the substitution of a different pyrimidine for the normal pyrimidine.

Translation - Protein synthesis; the process by which the genetic message carried by mRNA directs the synthesis of polypeptides with the aid of ribosomes and other cell constituents.

Trophozoite - The active, motile feeding stage of a protozoan organism; in the malarial parasite, the stage of schizogony between the ring stage and the schizont.

Tropism - The movement of living organisms toward or away from a focus of heat, light, or other stimulus.

Ultramicrobacteria - Bacteria that can exist normally in a miniaturized form or which are capable of miniaturization under low-nutrient conditions. They may be 0.2 m or smaller in diameter.

Ultraviolet (UV) radiation - Radiation of fairly short wavelength, about 10 to 400 nm, and high energy.

Uracil - The pyrimidine 2,4-dioxypyrimidine, which is found in nucleosides, nucleotides, and RNA.

V vaccine - A preparation of either killed microorganisms; living, weakened (attenuated) microorganisms; or inactivated bacterial toxins (toxoids). It is administered to induce development of the immune response and protect the individual against a pathogen or a toxin.

Vaccinomics - The application of genomics and bioinformatics to vaccine development.

Valence - The number of antigenic determinant sites on the surface of an antigen or the number of antigen-binding sites possessed by an antibody molecule.

Variable region (VL and VH) - The region at the N-terminal end of immunoglobulin heavy and light chains whose amino acid sequence varies between antibodies of different specificity. Variable regions form the antigen binding site.

Vasculitis - Inflammation of a blood vessel.

Vector - (1) In genetic engineering, another name for a cloning vector. A DNA molecule that can replicate (a replicon) and is used to transport a piece of inserted foreign DNA, such as a gene, into a recipient cell. It may be a plasmid, phage, cosmid or artificial chromosome. (2) In epidemiology, it is a living organism, usually an arthropod or other animal, that transfers an infective agent from one host to another.

Vector-borne transmission - The transmission of an infectious pathogen between hosts by means of a vector.

Vehicle - An inanimate substance or medium involved in the transmission of a pathogen.

Verrucae vulgaris (verruca vulgaris) - The common wart; a raised, epidermal lesion with horny surface caused by an infection with a human papillomavirus.

Vibrio - A rod-shaped bacterial cell that is curved to form a comma or an incomplete spiral.

Viral hemagglutination - The clumping or agglutination of red blood cells caused by some viruses.

Viral neutralization - An antibody-mediated process in which IgA, IgM, and IgA antibodies bind to some viruses during their extracellular phase and inactivate or neutralize them.

Viremia - The presence of viruses in the blood stream.

Viricide - An agent that inactivates viruses so that they cannot reproduce within host cells.

Virion - A complete virus particle that represents the extracellular phase of the virus life cycle; at the simplest, it consists of a protein capsid surrounding a single nucleic acid molecule.

Virioplankton - Viruses that occur in waters; high levels are found in marine and freshwater environments.

Viroid - An infectious agent of plants that is a single-stranded RNA not associated with any protein; the RNA does not code for any proteins and is not translated.

Virology - The branch of microbiology that is concerned with viruses and viral diseases.

Virulence - The degree or intensity of pathogenicity of an organism as indicated by case fatality rates and/or ability to invade host tissues and cause disease.

Virulence factor - A bacterial product, usually a protein or carbohydrate, that contributes to virulence or pathogenicity.

Virulent bacteriophages - Bacteriophages that lyse their host cells during the reproductive cycle.

Virus - An infectious agent having a simple acellular organization with a protein coat and a single type of nucleic acid, lacking independent metabolism, and reproducing only within living host cells.

Vitamin - An organic compound required by organisms in minute quantities for growth and reproduction because it cannot be synthesized by the organism; vitamins often serve as enzyme cofactors or parts of cofactors.

Wart (wort) - An epidermal tumor of viral origin.

Wastewater treatment The use of physical and biological processes to remove particulate and dissolved material from sewage and to control pathogens.

Water activity (aw) - A quantitative measure of water availability in the habitat; the water activity of a solution is one-hundredth its relative humidity.

Water mold - A common term for a member of the division *Oomycota.*

Weil-Felix reaction - A test for the diagnosis of typhus and certain other rickettsial diseases. In this test, the blood serum of a patient with suspected rickettsial disease is tested against certain strains of *Proteus vulgaris* (OX-2, OX-19, OX-K). The agglutination reactions, based on antigens common to both organisms, determine the presence and type of rickettsial infection.

White piedra - A fungal infection caused by the yeast *Trichosporon beigelii* that forms light-colored nodules on the beard and mustache.

Whole-genome shotgun sequencing - An approach to genome sequencing in which the complete genome is broken into random fragments, which are then individually sequenced. Finally the fragments are placed in the proper order using sophisticated computer programs.

Whole-organism vaccine - A vaccine made from complete pathogens, which can be of four types: inactivated viruses; attenuated viruses; killed microorganisms; and live, attenuated microorganisms.

Widal test - A test involving agglutination of typhoid bacilli when they are mixed with serum containing typhoid antibodies from an individual having typhoid fever; used to detect the presence of *Salmonella typhi* and *S.paratyphi.*

Winogradsky column - A glass column with an anaerobic lower zone and an aerobic upper zone, which allows growth of microorganisms under conditions similar to those found in a nutrient-rich lake.

Woolsorter's disease - *See* anthrax.

Wort - The filtrate of malted grains used as the substrate for the production of beer and ale by fermentation.

Xenograft - A tissue graft between animals of different species.

Xerophilic microorganisms - Microorganisms that grow best under low aw conditions, and may not be able to grow at high aw values

Yeast - A unicellular fungus that has a single nucleus and reproduces either asexually by budding or fission, or sexually through spore formation.

Yeast artificial chromosome (YAC) - A stretch of DNA that contains all the elements required to propagate a chromosome in yeast and which is used to clone foreign DNA fragments in yeast cells.

Yellow fever - An acute infectious disease caused by a flavivirus, which is transmitted to humans by mosquitoes. The liver is affected and the skin turns yellow in this disease.

YM shift - The change in shape by dimorphic fungi when they shift from the yeast (Y) form in the animal body to the mold or mycelial form (M) in the environment.

Zooflagellates - Flagellate protozoa that do not have chlorophyll and are either holozoic, saprozoic, or symbiotic.

Zoonosis - A disease of animals that can be transmitted to humans.

Zooplankton - A community of floating, aquatic, minute animals and nonphotosynthetic protists.

Zoospore - A motile, flagellated spore.

Zooxanthella - A dinoflagellate found living symbiotically within cnidarians and other invertebrates.

Zoster - *See* shingles.

***Z* value** - The increase in temperature required to reduce the decimal reduction time to one-tenth of its initial value.

Zygomycetes - A division of fungi that usually has a coenocytic mycelium with chitinous cell walls. Sexual reproduction normally involves the formation of zygospores. The group lacks motile spores.

Zygospore - A thick-walled, sexual, resting spore characteristic of the zygomycetous fungi.

Zygote - The diploid (2n) cell resulting from the fusion of male and female gametes.

History and Scope of Microbiology

Microorganisms are tiny creatures which can not be seen by the naked eye and can only be visualized under microscope

Branches of Microbiology

- Pure Microbiology (Taxonomic arrangement and Integrative arrangement)
- *Applied Microbiology.*

Taxonomic Arrangement

1. Bacteriology
2. Mycology
3. Phycology
4. Virology
5. Protozoology
6. Immunology

Integrative Arrangement

1. Microbial cytology
2. Microbial physiology
3. Microbial genetics
4. Microbial ecology
5. Microbial taxonomy
6. Cellular microbiology
7. Molecular microbiology

Applied Microbiology

1. Medical Microbiology
2. Veterinary Microbiology
3. Public Health Microbiology
4. Industrial Microbiology
5. Pharmaceutical Microbiology
6. Agriculture Microbiology
7. Plant Microbiology and Soil Microbiology
8. Food and Dairy Microbiology
9. Environmental Microbiology
10. Water/Aquatic Microbiology
11. Aero-microbiology
12. Microbial Biotechnology
13. Vaccinology
14. Chemotherapy

Naming Microorganisms

To identify all species of life on Earth Linnaeus – (1707-1778) Father of modern taxonomy Created Binomial nomenclature 2 names- Genus-species Names are italicized or underlined. The genus is capitalized and the specific epithet is lower case.

Names may be descriptive or honor a scientist: Staphylococcus aureus, Describes the clustered arrangement of the cells (staphylo) and the golden color of the colonies. Escherichia coli, Honors the discoverer, Theodor Eshcerich, and describes the bacterium's habitat, the large intestine or colon.

After the first use, scientific names may be abbreviated with the first letter of the genus and the specific epithet e.g. S. aureus.

Brief history of Microbiology

- Robert Hook (1665) – reported that life's smallest structural units of 'little boxes' or 'cells'. This marked the beginning of cell theory – that all living things are composed of cells.
- ***Cell Theory*** *In 1665, Robert Hooke reported that living things were composed of little boxes or cells. In 1858, Rudolf, Virchow said cells arise from preexisting cells.*
- *Cell Theory. All living things are composed of cells and come from preexisting cells*
- First observation of microbes 1673-1723, Antoni Van Leeuwenhoek described live microorganisms that he observed (Animalcules) Teeth scrapings Rain water Peppercorn infusions
- Antoni Van Leuwenhoek (1673)-discovered the 'invisible' world of microorganisms 'animalcules'.
- Until second half of nineteenth century many believed that some forms of life could arise spontaneously from non-living matter – spontaneous generation.
- Francesco Redi (1668)-Strong opponent of spontaneous generation. He demonstrated that maggots appear on decaying meat only when flies are able to lay eggs on the meat.
- John Needham (1745)-claimed that microorganisms could arise spontaneously from heated nutrient broth.
- Lazzaro Spallanzani (1765)-repeated Needhams experiments and suggested that Needham's results were due to microorganisms in the air entering the broth.
- Rudolf Virchow (1858)-concept of biogenesis – living cells can arise only from preexisting cells.
- Louis Pasteur (1822-1895)-Pasteur's experiments on swan shaped necks resolved the controversy of spontaneous generation. His discoveries led to the development of aseptic techniques used in the laboratory and medical procedure to prevent contamination by microorganisms that are in the air.

Golden age of microbiology

Rapid advances in the science of microbiology were made between 1857 and 1914.

Fermentation and Pasteurization

- Pasteur found that yeast ferments sugars to alcohols and that bacterium can oxidize the alcohol to acetic acid.
- Heating processes called pasteurization is used to kill bacteria in some alcoholic beverages and milk.

The Germ theory of disease

- Agostino Bassi (1934) and Pasteur (1865)-Showed a casual relationship between microorganisms and disease.
- Joseph Lister (1860s)-Introduced the use of disinfectant to clean surgical dressings in order to control infection in humans
- Robert Koch (1876)-Proved that microorganisms transmit disease-Koch's postulates which are used today to prove that a particular microorganism causes a particular disease.

Koch's postulates (Henle-Koch's Postulates) are

1. A specific organism should be found constantly in association with the disease.
2. The organism should be isolated and grown in a pure culture in the laboratory.
3. The pure culture when inoculated into a healthy susceptible animal should produce symptoms/lesions of the same disease
4. From the inoculated animal, the microorganism should be isolated in pure culture.
5. An additional criterion introduced is that specific antibodies to the causative organism should be demonstrable in patient's serum.

- Angelina – American wife of Koch's assistant suggested solidifying broths with agar as an aid to obtaining pure cultures.
- Koch also developed techniques for isolating organisms. Identified the bacillus that causes tuberculosis and anthrax, developed tuberculin and studied various diseases in Africa and Asia. His studies on Tuberculosis won him Nobel prize for philosophy and medicine in 1905.

Vaccination

- Immunity is conferred by inoculation with a vaccine.
- **Edward Jenner(1798)** – Demonstrated that inoculations with cowpox material provides humans with immunity from small pox
- **Pasteur (1880)** – Discovered that avirulent bacteria could be used as a vaccine for chicken cholera; he coined the word vaccine
- Modern vaccines are prepared from living avirulent microorganisms or killed pathogens, from isolated components of pathogens, and by recombinant DNA techniques.

Emergence of special fields of Microbiology

Immunology

- Immunization was first used against small pox. Edward Jenner used fluid from cowpox blisters to immunize against it.
- Pasteur developed techniques to weaken organisms so that they would produce immunity without producing disease.
- Elie Metchnikoff discovered that certain cells in the body would ingest microbes and named them as phagocytes

Industrial Microbiology and Microbial ecology

- **Pasteur** – Fermentation technology and pasteurization. One of his most important discovery was that some fermentative microorganisms were anaerobic and others were able to live either aerobically or anaerobically.

Microbial ecology – Two pioneers

- **Sergei N. Winogradsky (1856-1953)**-Soil microbiology-discovered that soil bacteria could oxidize Iron, Sulfur and Amonia to obtain energy and many bacteria incorporate CO_2 into organic matter. He also isolated anaerobic nitrogen fixing soil bacteria and studied the decomposition of cellulose.
- **Martinus Beijerinck (1851-1931)**-He isolated aerobic nitrogen fixing bacterium *Azotobacte*r, a root nodule bacterium also capable of fixing nitrogen (later renamed as *Rhizobium*); and sulfate reducing bacteria. Both of them developed enrichment culture technique and use of selective media, which have been of great importance in microbiology

Virology

- **Beijerinck** Characterized viruses as pathogenic molecules that could take over a host cells mechanisms for their own use
- **Wendell Stanley (1935)**-Crystallized TMV and crystals consisted of protein and RNA.
- **Viruses** were first observed with an EM in 1939.
- **Alfred Hershey and Martha Chase (1952)**-Demonstrated that the genetic material of some viruses is DNA
- James Watson and Francis Crick (1953)-Determined the structure of DNA

Chemotherapy

- There are two types of chemotherapeutic agents: synthetic drugs and antibiotics.
- **Elrlich (1910)**-Introduced an arsenic containing chemical called salvarsan to treat syphilis.

- **Alexander Fleming (1928)**-Observed that the mold *Penicillium* inhibited the growth of bacteria and named the active ingredient as penicillin. Penicillin has been used clinically as an antibiotic since the 1940s. Domagk and others developed sulfa drugs.
- **Waksman and others** developed Streptomycin and other antibiotics derived from soil organisms.
- Researchers are tackling the problem of drug-resistant microbes.

Genetics and Molecular Biology

- **1900**-Modern genetics began with the rediscovery of Gregor Mendel's principles of genetics.
- **Frederick Griffith (1928)**-Discovered that previously harmless bacteria could change their nature and become capable of causing disease
- **Avery, McCarty and MacLeod (1940's)**-Showed that this genetic change was due to DNA. After this finding came the crucial discovery of the structure of DNA by Watson and Crick
- **Edward Tatum and George Beadle**-Sstudied biochemical mutants of *Neurospora* to show how genetic information controls metabolism.
- **Barbara McClintock (1950)**-Discovered that some genes could move from one location to another on a chromosome.
- Early 1960's witnessed a further explosion of discoveries relating to the way DNA controls protein synthesis.
- **Francois Jacob and Jacques Monod (1961)**-Discovered mRNA and later made the first major discoveries about regulation of gene function in bacteria.
- Microorganisms can now be genetically engineered to manufacture large amounts of human hormones and other urgently needed medical substances.

- Late 1960's Paul Berg showed that fragments of human or animal DNA that code for important proteins can be attached to bacterial DNA. The resulting hybrid was the first example of recombinant DNA

Members of the microbial world: There are two types of cells

Prokaryotic cells having a simpler morphology and lack a true membrane bound nucleus. All bacteria are prokaryotic.

Eukaryotic cells have a membrane-bound nucleus; are more complex morphologically and larger than prokaryotes. Algae, fungi, protozoa, higher plants, and animals are eukaryotes

It is now clear that there are two quite different groups of prokaryotic organisms; Bacteria and Archaea.

Microbiologists proposed that organisms should be divided among three domains; Bacteria (the true bacteria or eubacteria), Archaea and Eucarya (all eukaryotic organisms).

Archaea

The Archaea are a gathering of single-celled microorganisms. They have no cell core or some other layer bound organelles inside their cells. Archaea and microscopic organisms are very comparable fit as a fiddle, in spite of the fact that a couple of archaea have extremely unordinary shapes, for example, the level and square-formed cells. Likeness to microscopic organisms, archaea have qualities and a few metabolic pathways that are all the more firmly identified with those of eukaryotes, outstandingly the chemicals engaged with interpretation. The archaea organic chemistry is one of a kind, for example, nearness of ether lipids in their cell films. Archaea utilize a much more prominent assortment of wellsprings of vitality than eukaryotes: going from natural mixes, for example, sugars, to alkali, metal particles or even Hydrogen gas. Archaea reproduce asexually by binary fission, fragmentation, or budding; unlike bacteria and eukaryotes, no known species form spores. At first archaea were viewed as extremophiles that lived in brutal conditions, for example, hot springs and salt lakes, yet they are currently found in a wide scope of natural surroundings, including soils, seas, marshland. Archaea assume jobs in both the carbon cycle and the nitrogen cycle. No archaea pathogens or parasites are known, yet they are regularly mutualists or commensals. Methanogens are utilized in biogas generation and sewage treatment, and compounds from extremophile archaea that can bear high temperatures and natural solvents are abused in biotechnology.

Bacteria

Microscopic organisms are an extensive space of prokaryotic-microorganisms. Microscopic organisms are available in most living spaces on Earth, developing in soil, acidic hot springs, radioactive waste water, natural issue and live assemblages of plants and creatures. Microscopic organisms have numerous shapes and sizes. Bacterial cells are around one tenth the extent of eukaryotic cells and 0.5-5.0 micrometers long. Most bacterial species are either circular, called cocci or rodshaped, called bacilli. Some bar formed microbes are marginally bended called vibrio or comma-molded. Numerous bacterial species exist as single cells and partner in trademark examples, for example, frame sets called diploids, shape chains, and gathering together in groups. Microscopic organisms can likewise be prolonged to frame fibers. The bacterial cell is encompassed by cell layer, which encases the substance of the cell also, goes about as a boundary to hold supplements, proteins and other fundamental segments of the cytoplasm inside the cell. They come up short on a genuine core, mitochondria, chloroplasts, Golgi complex and endoplasmic reticulum. Most microscopic organisms don't have a layer bound core, and their hereditary material is normally a solitary roundabout chromosome situated in the cytoplasm in a sporadically molded body called the nucleoid.The nucleoid contains the chromosome with related proteins and RNA. The microscopic organisms contain ribosomes for the generation of proteins but unique in relation to those of eukaryotes and Archaea. A few microscopic organisms create intracellular supplement stockpiling granules, for example, glycogen, polyphosphate, sulfur polyhydroxyalkanoates.These granules empower microorganisms to store mixes for some other time utilize. Certain bacterial species, for example, the photo-synthetic Cyanobacteria deliver inward gas vesicles which they use to manage their lightness-enabling them to climb or down into water layers with various light forces and supplement levels. The cell divider is present outwardly of the cytoplasmic layer. A typical bacterial cell divider material is peptidoglycan which is a polymer contains two sugar subordinates Naetylglucosamine(NAG) and N-acetylmuramic corrosive (NAM) joined by glycosidic bond. A peptide chain of four substituting D- and L-amino acids called tetrapeptide is associated with the carboxyl gathering of the NAM. The amino acids present in the tetrapeptide incorporate Lalanine, D-alanine, D-glutamic corrosive, and either lysine or diaminopimilic corrosive (DAP).The carboxyl gathering of terminal D-alanine is associated straightforwardly to amino gathering of DAD. The peptide interbridge interfaces the tetrapeptide chains. There are two distinct sorts of cell divider in microorganisms, called Gram-positive and Gram-negative. Gram-positive microbes have a thick cell divider containing numerous layers of peptidoglycan and teichoic acids. Conversely, Gram-negative microorganisms have a generally thin cell divider comprising of a couple of layers of peptidoglycan encompassed by a second lipid film containinlipopolysaccharides and lipoproteins. They have many surface structures, for example, flagella, pili and fimbriae. Flagella are inflexible protein structures about 20 nanometers in breadth and up to 20 micrometers long that are utilized for motility. Flagella are driven by the vitality discharged by the exchange of particles down an electrochemical inclination over the cell film. Fimbriae are fine fibers of protein, just 2-10 nanometers in breadth and up to a few micrometers long. They are disseminated over the surface of the cell, and look like fine hairs. Fimbriae are associated with connection to strong surfaces or to different cells and are basic for the harmfulness of some bacterial pathogens.Pili are cell limbs, somewhat bigger than fimbriae that can exchange hereditary material between bacterial cells in a procedure called conjugation. Containers or sludge layers are created by microorganisms to encompass their cells, and shift in auxiliary unpredictability for example, disorganized slime layer and exceedingly organized container or glycocalyx. These structures shield cells from engulfment by eukaryotic cells, for example, macrophages. They can likewise go about as antigens and be engaged with cell acknowledgment. Gram-positive microorganisms, such as *Bacillus, Clostridium, Sporohalobacter, Anaerobacter* and *Heliobacterium*, can shape exceptionally safe, lethargic structures called endospores. Endospores have cytoplasm containing DNA and ribosomes encompassed by a cortex layer and ensured by animpermeable and unbending coat. Endospores can endure extraordinary physical and compound stresses, for example, elevated amounts of UV light, gamma radiation, cleansers, disinfectants, warm, solidifying, weight and drying up in this manner help in getting by in cruel conditions. The microscopic organisms for the most part imitated by paired splitting which includes chromosome replication pursued by cell division. Yet, microorganisms recombine their hereditary materials by three different ways:-

1. Conjugation
2. Transformation
3. Transduction

The bacteria are classified based on their source of energy, carbon and hydrogen/ electron source are of following type:

Based on carbon source: Autotrophs whose main carbon source is carbon dioxide and heterotrophs whose carbon source is reduced organic molecules.

Based on energy source: Phototrophs are the microorganisms which uses the light as their energy source and chemotrophs who get their energy by oxidation of organic and inorganic compounds.

Based on hydrogen and electron source: Lithotrophsar the microorganisms in which the electron source is reduced inorganic molecule and organotrophs whose electron source is organic molecules.

The bacterial growth mainly involves increase in cell mass and cell division. Under favorable condition bacteria grow in geometric progression i.e. doubles at regular intervals. This growth is called exponential growth. The bacterial growth can be divided into four phases as:

Lag phase: The population remains temporarily unchanged and no apparent cell division though cell may be growing in volume and mass.

Log phase: Where the cells are dividing regularly by binary fission and growing by geometric progression. The cells divide at constant rate based on growth medium.

Stationary phase: The population growth is limited due to nutrients exhaustion, accumulation of inhibitory metabolites or end products and limitation of biological spaces.

Death phase: Due to limitation of nutrients bacteria die and no more cell divisions.

The generation time of bacteria and growth rate can be calculated from the growth curve by the equation: G (generation time) = (time, in minutes or hours)/n (number of generations).

Algae

They are photosynthetic eukaryotes. They have diverse sorts of photosynthetic Pigments i.e. chlorophyll. They are for the most part found in damp condition. They are minuscule and buoy in surface waters (phytoplankton) and live appended to rough drifts (kelp). Measure ranges from 0.5 μm to more than 50 m long. Need vascular tissues-no evident roots, stems, or clears out. They chiefly replicated by both sexual and agamic methods of multiplication and have no multicellular conceptive organs. They are a wide range of kinds of green growth, for example,

Red algae: Their size and multifaceted nature shift from thin movies developing on rocks to complex filaments.Their frill colors called phycobilins veil the chlorophyll and give them their red shading. Because of these particular shades, red green growth are frequently ready to photosynthesize in more profound water than other green growth. Red green growth don't have flagella.They have numerous advantages such as utilized as sustenance and research center item i.e. agar used to develop microorganisms and parasites is gotten from red green growth.

Green algae

They are found for the most part in crisp waters and ashore. Most species coast in streams, lakes, supplies, and springs. They can likewise live on rocks, soil, and tree covering. Green growth are living beings with an assortment of body frames including single cells, fibers, provinces, and thalli. They have the equivalent photosynthetic shades (chlorophyll an and b) what's more, some green growth have firm cell dividers made out of cellulose, as do plants.

Dinoflagellates

Found in warm, Tropical Ocean. They are essentially unicellular. Green and drab structures, phagotrophic and parasitic. They are biflagellate. Their nucleus is irregular. Some are bioluminescent structures light up when water is bothered and they generally replicated by agamically. Dark colored green growth: They are known as chilly water green growth and found in rough drift in calm zone or open

Ocean. Most dark colored green growth contain the shade fucoxanthin, which is in charge of the particular greenish-darker shading. They are multicellular and duplicated by whipped spores.

Diatoms

They are most normal kinds of phytoplankton and furthermore known as brilliant dark colored green growth. They are generally unicellular and can exist as settlements in the state of fibers or strips. Their divider made of silica called a frustule. They are usually utilized in investigations of water quality. A few diatoms are fit for development through whipping. They imitated by abiogenetic for a few ages, at that point sexual.

Fungi

They are eukaryotic living beings that include yeasts, molds and mushrooms. They are nonphotosynthetic and contain no chlorophyll shades. A large portion of them are multicellular and some are unicellular e.g. yeast. They are non motile and need genuine leaves, roots and stems. Organisms need warm, wet spots to develop. They are found essentially in clammy sustenances, sodden tree barks, and wet restroom tiles etc.Fungi are heterotrophs that feed by retention. They ingest little natural particles from the encompassing medium. The chemicals and hydrolytic chemicals emitted by the growth separate sustenance outside its body into more straight forward exacerbates that the parasite can retain and utilize. The absorptive method of nourishment is related with the natural jobs of growths as decomposers, parasites, and mutualistic symbionts. Saprobic parasites assimilate supplements from nonliving creatures. Parasitic organisms assimilate supplements from the cells of living hosts. The parasitic cells contain membrane bound nuclei with chromosomes that contain DNA with noncoding locales called introns further more, coding locales called exons. They additionally have film bound cytoplasmic organelles, for example, mitochondria, sterol-containing layers, and ribosomes of the 80S type. They have dissolvable starches and capacity mixes, including sugar alcohols, disaccharides, and polysaccharides. Growths need chloroplasts and are heterotrophic life forms, requiring preformed natural mixes as vitality sources. Organisms have a cell divider and vacuoles. They recreate by both sexual and abiogenetic means and produce spores. They have haploid cores. The cells of most organisms develop as tubular, prolonged, and string like structures are called hyphae which may contain various cores. A few animal categories develop as single-celled yeasts that repeat by sprouting or parallel parting. The parasitic cell divider is made out of glucans and chitin. Most parasites develop as hyphae which are barrel shaped, string like structures 2-10 μm in breadth & up to a few centimeters long. Hyphae develop at their tips; new hyphae are regularly framed by a procedure called expanding, or developing hyphal tips bifurcate offering ascend to two parallel-developing hyphae. Hyphae can be either septate or coenocytic: septate hyphae are isolated into compartments isolated by cross dividers, with every compartment containing at least one cores; coenocytic hyphae are not compartmentalized. Contagious propagation is intricate. They imitated by both explicitly and abiogenetically. Abiogenetic generation through vegetative spores (conidia) or through mycelial fracture. Sexual propagation includes joining of hyphae is called conjugation, two mating strains with different nuclei shape consistent film is known as plasmogamy and some of the time the nuclei are combined is called karyogamy.

Protozoa

Protozoa are parasitic and creature like protists in view of their motility. Their sizes extend from 10 to 52 micrometers. They moved by flagella, hair-like structures called cilia and foot-like structures called pseudopodia. Protozoa retain sustenance by their cell layers e.g., single adaptable cells, encompass nourishment and inundate it. All protozoa process their sustenance in stomach-like compartments called vacuoles. Protozoa can duplicate by paired parting or numerous parting. Some protozoa replicate explicitly, some abiogenetically, while some utilization a mix. They cause numerous infections in human, for example, Malaria, Amoebiasis and Leishmaniasis and so forth.

The Future of Microbiology

As the first segments have appeared, microbiology has had a significant effect on society. What of things to come? Science essayist Bernard Dixon is exceptionally hopeful about microbiology's future for two reasons. To begin with, microbiology has a clearer mission than do numerous other logical orders. Second, it is sure of its esteem as a result of its viable criticalness. Dixon noticed that microbiology is required both to confront the danger of new and reemerging human irresistible sicknesses and to create modern innovations that are more effective and ecologically amicable. What are probably the most encouraging zones for future microbiological inquire about and their potential viable effects? What sorts of difficulties do microbiologists confront? The accompanying brief rundown should give some thought of what the future may hold: 1. New irresistible illnesses are consistently emerging and old illnesses are by and by getting to be across the board and damaging. Helps, hemorrhagic fevers, and tuberculosis are magnificent precedents of new and reemerging irresistible illnesses. Microbiologists should react to these dangers, huge numbers of them by and by obscure. 2. Microbiologists must discover approaches to stop the spread of set up irresistible maladies. Increments in anti-microbial opposition will be a proceeding with issue, especially the spread of various medication obstruction that can render a pathogen impenetrable to current restorative treatment. Microbiologists need to make new medications and discover approaches to moderate or keep the spread of medication obstruction. New antibodies must be produced to ensure against infections, for example, AIDS. It will be important to utilize procedures in sub-atomic science what's more, recombinant DNA innovation to tackle these issues. 3. Research is required on the relationship between irresistible operators and ceaseless infections, for example, immune system and cardiovascular ailments. It might be that a portion of these incessant tribulations mostly result from diseases. 4. We are just presently starting to see how pathogens connect with host cells and the manners by which ailments emerge. There additionally is a lot to find out about how the host stands up to pathogen attacks. 5. Microorganisms are progressively essential in industry and natural control, and we should figure out how to utilize them in an assortment of new ways. For instance, microorganisms can (a) fill in as well springs of excellent nourishment and other useful items, for example, catalysts for modern applications, (b) debase toxins and harmful squanders, and (c) be utilized as vectors to treat infections and upgrade agrarian efficiency. There likewise is a proceeding with need to ensure nourishment furthermore, crops from microbial harm. Microbial assorted variety is another region requiring impressive inquire about. Undoubtedly, it is evaluated that under 1% of the earth's microbial populace has been refined. We should grow new disconnection procedures and a satisfactory characterization of microorganisms, one which incorporates those microorganisms that can't be developed in the lab. Much work should be done on microorganisms living in outrageous conditions. The disclosure of new microorganisms may well prompt further advances in modern procedures and improved natural control. 7. Microbial people group frequently live in biofilms, and these biofilms are of significant significance in both medication and microbial nature. Research on biofilms is in its early stages; it will be numerous prior years we all the more completely comprehend their nature and can utilize our insight in commonsense ways. By and large, organism microorganism communications have not yet been widely investigated. 8. The genomes of numerous microorganisms as of now have been sequenced, and numerous more will be resolved in the coming years. These groupings are perfect for figuring out how the genome is identified with cell structure and what the least variety of qualities important. Investigation of the genome and its movement will require proceeding with advances in the field of bioinformatics and the utilization of PCs to examine organic issues. 9. Additionally inquire about on abnormal microorganisms and microbial biology will prompt a superior comprehension of the communications among microorganisms and the lifeless world. In addition to other things, this comprehension ought to empower us to all the more successfully control contamination. So also, it has turned out to be evident that microorganisms are fundamental accomplices with higher living beings in advantageous connections. More prominent learning of advantageous connections can help enhance our valuation for the living scene. It likewise will prompt upgrades in the soundness of plants, domesticated animals, and people. 10. Due to their relative effortlessness, microorganisms are astounding subjects for the investigation of an assortment of crucial inquiries in science. For instance, how do complex cell structures create and how do cells speak with one another and react to the earth? 11. At last, microbiologists will be tested to precisely survey the ramifications of new disclosures and mechanical improvements. They should convey a decent perspective of both the positive and negative long haul effects of these occasions on society. The eventual fate of microbiology is splendid. The microbiologist René Dubos has condensed well the energy and guarantee of microbiology.

Vaccine Recommendations for Travelers Aged 2 Years or Older

For trevellers who are 2 years or more following immunization is recommended

S No.	Vaccine Name	Dose	Age at which vaccine should be administered
1	Measles, mumps, and rubella (MMR) vaccine	At least 1 dose	on or after 12 months of age
2	Diphtheria, tetanus, and acellular pertussis (DtaP)	4 or 5 doses	until age 7 after age 7, 1 dose of adult tetanus and diphtheria (Td) vaccine every 10 years
3	Polio vaccine	At least 3 doses	Upto 5 years
4	Haemophilus influenzae type b (Hib) vaccine	1	not recommended after 5 years of age
5	Hepatitis B vaccine	3 doses	At any age
6	Varicella vaccine (for persons who have never had chickenpox)	3-4	12 and 15 months old
7	Influenza (flu) vaccine	2	for adults 65 years or older and for other high-risk individuals
8	Pneumococcal vaccine	1	recommended for adults 65 years or older and for other high-risk individuals

27 University/Colleges offering courses in Microbiology

1. Aligarh Muslim University, Aligarh (Uttar Pradesh)
2. All India Institute of Medical Science (AIIMS), Delhi
3. Assam University, Silchar (Assam)
4. AVK Group of Institute, Bangalore
5. Baba Farid Institute of Technology
6. Bangalore City College
7. Bangalore City College, Bangalore
8. BBAU-Babasaheb Bhimrao Ambedkar University, Lucknow
9. Bharathidasan University, Tiruchirappalli (Tamil Nadu)
10. Brabourne College, Kolkata
11. Brindavan College, Bangalore
12. Chaudhary Charan Singh Haryana Agricultural University, Hisar (Haryana)
13. Christian Medical College, Vellore (Tamil Nadu)
14. CVRU - Dr. C.V. Raman University, Chhattisgarh
15. Dr. Babasaheb Ambedkar Marathwada University, Mumbai
16. Dr. D.Y. Patil University, Mumbai
17. East West Group of Institutions, Bangalore
18. Elective Government College for Girls, Chandigarh
19. Ethiraj College for Women, Chennai
20. Faculty of Science & Faculty of Life Sciences, Aurangabad (Andhra Pradesh)
21. Fergusson College, Pune (Maharashtra)
22. Ganga Kaveri group of Institutions, Bangalore
23. Gargi college, Delhi
24. Gayatri College of Biomedical Science, Dehradun

25. Goa University : Faculty of Science, Taleigao Plateau (Goa)
26. Gujarat University : Faculty of Science, Ahmedabad (Gujarat)
27. Guru Nanak Dev University, Amritsar (Punjab)
28. Himachal Pradesh Krishi Vishwavidyalaya, Kangra (Himachal Pradesh)
29. Indian Academy Group of Institutions, Bangalore
30. Jawaharlal Institute of Postgraduate Medical Education and Research, Pondicherry
31. Jiwaji University, Faculty of Sciences & Faculty of Life Sciences, Vidya Vihar, Gwalior
32. Kakatiya University, Vidyaranyapuri, Warangal
33. Karnatak University, Dharwad (Karnataka)
34. KCCACS, Mumbai
35. Kristu jAyanti college of Management and technology, Bangalore
36. Kurukshetra University, Faculty of Science, Kurukshetra
37. Lady Brabourne College, Kolkata
38. Lalit Narayan Mithila University, Darbhanga (Bihar)
39. Madras Christian college ,chennai
40. Madurai Kamaraj University - Madurai
41. Maharshi Dayanand University, Faculty of Science, Rohtak
42. Mahatma Gandhi University : Faculty of Applied Science, Kottayam (Kerala)
43. Manipal Academy of Higher Education,Manipal (Karnataka)
44. Manonmaniam Sundaranar University, University Campus, Abishekapatti, Tirunelveli
45. Mithibai College, Delhi
46. Mumtaz College, Hyderabad
47. Mysore University - UOM (University of Mysore)

48. North Maharashtra University, Faculty of Science, Umavinagar, Jalgaon
49. Orissa University of Agriculture & Technology, Bhubaneshwar (Orissa)
50. Osmania University, Hyderabad
51. Panjab University, Sector-14, Chandigarh
52. Pt. Ravishankar Shukla University, Faculty of Life Science, Raipur
53. Punjab University, Chandigarh
54. Sophia Institute of Medical Sciences, Gwalior
55. SRMCRI - Sri Ramachandra Institute of Higher Education and Research, Porur, Chennai
56. St Xavier's College
57. St. Francis College for Women, Hyderabad
58. St. Joseph's College, Bangalore
59. St.George College of Management, Science & Nursing, Bangalore
60. The Oxford Educational Institute, Bangalore
61. Topiwala National Medical college, Mumbai
62. University of Delhi, Delhi
63. University of Hyderabad, Hyderabad
64. University of Mumbai, Mumbai

Research Centers where microbiology related work is going on

1. Central Food Technological Research Institute, Mysore, Karnataka (www.cftri.com/)
2. Central Institute of Fisheries Technology, Willingdon Island, Kochi, Kerala

3. Centre for Scientific Novel therapeutics from plants & Prof. (Dr.)
4. Centre of Advanced Study in Botany, Department of Botany, Banaras Hindu University, Varanasi, Uttar Pradesh (http://www.bhu.ac.in/)
5. Department of Microbiology, Microbial enzyme technology; food biotechnology; health care ,
6. Department of Pharmaceutical Technology, Jadavpur University, Kolkata, West Bengal (www.jaduniv.edu.in/)
7. Faculty of Arts, Science and Humanities, Karpagam University, Coimbatore
8. CSIR-IMTECH, Chandigarh
9. ICAR-Indian Agricultural Research Institute, New Delhi
10. Indian Institute of Crop Processing Technology, Thanjavur, Tamil Nadu(http://www.iicpt.edu.in/)
11. ICAR-Indian Institute of Pulses Research (IIPR) Kanpur, Uttar Pradesh (http://www.iipr.res.in/)
12. Institute of Forest Genetics and Tree Breeding, Tamil Nadu, India
13. Kakatiya University, Department of Biotechnology and Department of Botany, Warangal, Andhra Pradesh (www.kakatiya.ac.in)
14. MACS Agarkar Research Institute, Pune
15. ICAR-National Dairy Research Institute,Kernal Haryana
16. ICAR-NBAIM, Mau Nath Bhanjan, U.P.
17. People's University, Bhopal (M.P.) [www.peoplesuniversity.ed u.in]
18. School of Bio Sciences and Technology, VIT University, Vellore, Tamilnadu (http://www.vit.ac.in)
19. Tamil Nadu School of Pharmacy & Research, (Formerly People's Institute of Pharmacy and Research Center)

International Depositary Authority (IDA) for microbial culture

S.No.	Country / Territory	International Depositary Authority
1.	Hungary	National Collection of Agricultural and Industrial Microorganisms (NCAIM)
2.	India	Microbial Type Culture Collection (MTCC)
3.	India	Microbial Type Culture Collection and Gene Bank (MTCC)
4.	Italy	Advanced Biotechnology Center (ABC)
5.	Australia	The National Measurement Institute (NMI)
6.	Bulgaria	National Bank for Industrial Microorganisms an Cell Cultures (NBIMCC)
7.	Canada	International Depository Authority of Canada (IDAC)
8.	Chile	Colección Chilena de Recursos Genéticos Microbianos (CChRGM)
9.	China	China Center for Type Culture Collection (CCTCC)
10.	Czech Republic	Czech Collection of Microorganisms (CCM)
11.	Finland	VTT Culture Collection (VTTCC)
12.	France	Collection nationale de cultures de micro-organismes (CNCM)
13.	Germany	Leibniz-Institut DSMZ - Deutsche Sammlung von Mikroorganismen und Zellkulturen GmbH (DSMZ)
14.	Hungary	National Collection of Agricultural and Industrial Microorganisms (NCAIM)
15.	Japan	National Institute of Technology and Evaluation, Patent Microorganisms Depositary (NPMD)
16.	Latvia	Microbial Strain Collection of Latvia (MSCL
17.	Mexico	Colección de Microorganismos del CNRG (CM-CNRG)
18.	Morocco	Collections Coordonnées Marocaines de Microorganismes (CCMM)
19.	Netherlands	Westerdijk Fungal Biodiversity Institute (CBS)
20.	Poland	IAFB Collection of Industrial Microorganisms Institute of Agricultural and Food Biotechnology (IAFB)
21.	Poland	Polish Collection of Microorganisms (PCM)
22.	Republic of Korea	Korean Agriculture

23.	Russian Federation	Russian Collection of Microorganisms (VKM)
24.	Slovakia	Culture Collection of Yeasts (CCY)
25.	Spain	Banco Español de Algas (BEA)
26.	Spain	Colección Española de Cultivos Tipo (CECT)
27.	Switzerland	Culture Collection of Switzerland (CCOS)
28.	United Kingdom	CABI Bioscience, UK Centre (IMI)
29.	United Kingdom	National Collections of Industrial, Food and Marine Bacteria (NCIMB)
30.	United States of America	American type culture collection (ATCC)

Unit-I

Soil Microbiology

Q1. Father of Soil Microbiology?

a) Antony van Leeuwenhoek b) M. Beijerinck

c) Louis Pasteur d) Sergei Winogradsky

Q2. Name the bacteria that prey on other bacteria?

a) *Nitrobacter*

b) *Bdellovibrio bacteriovorus*

c) *Proteobacteria*

d) *Pesudomonas*

Q3. What is geosimn?

a) Geosmin an compound found in soil for responsible for brown color of soil

b) Geosmin, an earthy-smelling substance, has been isolated from several actinomycete

c) Both A and B are correct

d) None of the above

Q4. Which of the following is correct?

a) Mycorrhizae are fungi that form a mutually beneficial (symbiotic) relationship with plant roots

b) The fungi aid in transmitting nutrients and water to the plant roots

c) The increased nutrient availability from mycorrhizae is thought to be due to the additional absorbing surface provided by the fungi

d) All of the above

Q5. The transformation of nitrates to gaseous nitrogen is accomplished by microorganisms in a series of biochemical reactions. The process is known as

a) Ammonification
b) Nitrogen fixation
c) Denitrification
d) Nitrification

Q6. The root nodule formation and development is monitor by which one of

a) *Bradyrhizobium*
b) *E. coli*
c) *Xanthomaonas*
d) None of the above

Q7. Which one of the following is responsible for nod factor in bacteria?

a) Fix gene
b) Gag gene
c) Nif gene
d) Nod genes

Q8. Which one of the following structure is formed in plant roots by mycorrhize?

a) Arbuscles
b) Hartig net
c) Haustoria
d) Rhizomorph

Q9. Actinomycetes prefers which pH for their growth?

a) 7 to 8.5
b) 4.5 to 6.7
c) 6.5 to 8.0
d) Only at 7

Q10. Which one of the following fungal group form ectomycorrhizae with the plant roots??

a) Ascomycota
b) Basidiomycota
c) Both a and b
d) None of the above

Q11. The diagnostic enzyme for nitrogen-fixing organisms is

a) NIitrogenase b) Nitrate reductase

c) Nitrate oxidase d) None of the above

Q12. Which of the following bacterium is called as the superbug that could clean-up oil spills?

a) *Bacillus subtilis* b) *Pseudomonas pudita*

c) *Bacillus denitrificans* d) *Pseudomonas denitrificans*

Q13. The process of converting environmental pollutants into harmless product by naturally occurring microbes is called

a) Exsitu bioremediation b) Intrinsic bioremediation

c) Extrinsic bioremediation d) None of these

Q14. Phosphorus and other nutrient to uptake is improved by

a) *Saccharomyces cerevisiae* b) *VA Mycorrhiza*

c) *Candida torulopsis* d) *Aspergillus niger*

Q15. The energy value of biogas is typically?

a) 400-700 BTU/ft^3 b) 1,000 BTU/ft^3

c) 1500 BTU/ft^3 d) more than 5000 BTU/ft^3

Q16. Degree of compost maturity can be assesed by

a) Infrared technique b) Germination test

c) Both (a) and (b) d) None of the above

Q17. Nitrification was discovered to be a biological process by

a) Sergei Winogradsky b) Selman Waksman

c) Ernst Chain d) None of these

Q18. Denitrification is carried out?

a) Usually by facultative anaerobes

b) Predominantly by *Pseudomonas* spp

c) Predominantly by *Bacillus* spp

d) All of the above

Q19. For rapid decomposition by microbes, the substrate should have a C/N ratio of

a) 10-20 b) 20-30

c) 30-40 d) 60- 80

Q20. Microorganism remove metals by?

a) Adsorption and complextion

b) Adsorption and precipitation

c) Adsorption and volatilization

d) All of these

Q21. Chlorella sps. are widely used for the removal of ?

a) Plastic b) Hydrocarbons

c) Organic matter d) Heavy metal

Q22. Which are the main source of biofertilisers?

a) Cyanobacteria b) Bacillus

c) Streptococcus d) None of these

Q23. Ex- situ bioremediation involves the ?

a) Degradation of pollutants by microbe directly

b) Removal of pollutant and collection at a place to facillate microbial degradtion

c) Degradation of pollutant by genetically engineered microbes

d) None of these

Q24. Which of the following microbe is widely used in the removal of industrial waste?

a) *Trichoderma* sps b) *Aspergillus niger*

c) *Pseudomonas putida* d) All of these

Q25. The process of extracting metal from ore bearing rocks is called?

a) Biodegradation b) Bioremediation

c) Both a and b d) Bioleaching

Q26. Which of the following types of association is present among Staphylococcus aureus and Aspergillus terreus?

a) Antagonism b) Mutualism

c) Parasitism d) Commensalism

Q27. Name the non symbiotic nitrogen fixing bacteria?

a) Frankia b) Azotobacter

c) Cyanobacteria d) Rhizobia

Q28. Give an example of symbiotic nitrogen fixer root nodule formed by a member of Actinomycetes?

a) *Frankia* b) *Micromonospora*

c) *Actinoplanes* d) *Nocardia*

Q29. The phenomenon of commensalism refers to a relationship between organisms in which?

a) One species of a pair benefits

b) Both the species of a pair benefit

c) One species of a pair is more benefited

d) None of the above

Q30. Name the soil bacterium which is a pathogen insects?

a) *Bacillus subtilis* b) *Bacillus thuringiensis*

c) *Pseudomonas* spp. d) A and B

Q31. Which of the following conditions decreases the level of denitrification?

a) Abundance of organic matter

b) Acidic pH

c) Elevated temperatures

d) Availability of oxygen

Q32. Agrobacterium is involved in which of the following processes?

a) Ammonification

b) Nitrification

c) Reduction of nitrate to ammonia

d) Denitrification

Q33. Nitrosococcus nitrosus is a nitrite-oxidizing bacteria.

a) True b) False

Q34. Aroma of earthy smell in first shower of monsoon is due to?

a) Geosmin b) Organic compounds

c) Algae d) None of the above

Q35. Which group of microbes found maximum in soil ?

a) Bacteria b) Fungi

c) Algae d) Protozoa

Q36. Ammonia oxidizers and nitrite oxidizers are

a) Gram-negative chemolithotrophs

b) Gram-positive chemolithotrophs

c) Gram-negative photolithotrophs

d) Gram-positive photolithotrophs

Q37. The Winogradsky column experiment is done in the dark.

a) True b) False

Q38. Which among the following are the most important agents for carbon dioxide fixation?

a) Bacteria b) Fungi

c) Algae d) Protozoa

Q39. Cellulose is degraded to cellobiose by the enzyme _______

a) Cellulase b) Beta-glucosidase

c) Hexokinase d) Cellulose dehydrogenase

Q40. Which of the following are not the features of component II of nitrogenase enzyme complex?

a) Component II is nitrogenase reductase

b) Component II is known as the MoFe protein

c) Contains sulfur

d) Not active without component I

Q41. Which among the following develop in the upper portion of the Winogradsky column?

a) Sulfate-reducing bacteria b) Green-sulfur bacteria

c) Purple-sulfur bacteria d) *Thiobacilli*

Q42. Which among the following is a non-sulfur purple bacteria?

a) *Rhodomicrobium* b) *Thiobacillus*

c) *Chromatium* d) *Chlorobium*

Q43. When two different Sps. of microorganisms occupy the same environment without affecting each other is ?

a) Commeansalisim b) Neturalism

c) Symbiosis d) Paratisim

Q44. A type of Association that involves the exchange of nutrient B/w the two sps a phenomenon is called?

a) Mutualism b) Syntrophism

c) Commensalism d) Antagonism

Q45. Which one of the following microorganisms play a key role in the transformation of rock to soil , first step in rock plant sucession?

a) Actinomycetes b) Cyanobacteria

c) Sulfur bacteria d) Molds

Q46. Which one of the following statement are true, except?

a) *Clostridium pasteurianum* is anaerobic nitrogen fixing

b) Nitrification process in soil is contributions of winogradsky

c) Louis paster is a father of soil microbiology

d) Some bacteria are good source of antibiotcs

Q47. Which relationship is this in which one sps. of pair benefits and the other is not affected?

a) Commensalism b) Amensalism

c) Anatagonism d) Mutulisim

Q48. The conversion of nitrogen to ammonia or nitrogenous compoundis called as?

a) Nitrification b) Nitrogen fixation

c) Nitrogen assimilation d) None of the above

Q49. The root nodule of legumes contains pink pigment which has high affinity for oxygen is?

a) Haemoglobin b) Bacterail haemoglobin

c) Nodhaemoglobin d) Leghaemoglobin

Q50. Plant absorbs N2 in the form of ?

a) Nitrites b) Nitrates

c) Ammonium d) All the above

Q51. To fix one molecule of nitrogen

a) 1 ATP molecule are required

b) 12 ATP molecule are required

c) 16 ATP molecule are required

d) 6 ATP molecule are required

Q52. Anabaena, a nitrogen fixer is present in a root packet of?

a) Azolla b) Salvinia

c) Pistia d) Marselia

Q53. Plant cannot absorb molecular nitrogen in the atmoshphere because?

a) N2 has double bonds making its highly stable

b) Abudance in the atmoshphere inhibits the absorption

c) N2 has triple bonds making its highly stable

d) None of these

Q54. Which of the following bacterial genus is capable of oxidizing ammonia (NH_4)?

a) Nitrospina b) Nitrobacter

c) Nitrococcus d) Nitrosobacter

Q55. The dominant mineral particles in most soils are compounds of ____________

a) Sodium b) Potassium

c) Magnesium d) Iron

Q56. Which of the followings is an anaerobic bacterium?

a) Azotobacter b) Nitrobactor

c) Clostridium d) None

Q57. Which of the following microorganism fixed nitrogen in waterlogged soil?

a) Nostoc b) Nitrobactor

c) Clostridium d) None

Q58. The root nodule forming bacteria are?

a) Azotobacter b) Nitrobacter

c) Clostridium d) Rhizobium

Q59. Which of the following organisms are known to grow on the surfaces of freshly exposed rocks?

a) Green algae b) Diatoms

c) Cyanobacteria d) Yeast

Q60. The major enzyme involved in biological nitrogen fixation are

a) Nitrogenase and hexokinase

b) Nitrogenase and hydrogenase

c) Nitrogenase and hydrolyase

d) Nitrogenase and peptidase

Q61. An example of ammonification is

a) Converting nitrogen into ammonia by nitrogen fixers

b) Converting urea into ammonia by decomposers

c) Converting nitrates into ammonia by nitrogen fixers

d) Converting nitrates into ammonia by decomposers

Q62. Majority of nitrogen fixation occurs by a?

a) Biological nitrogen fixing organisms

b) Lightning

c) Volcanic eruptions

d) Haber – bosch process

Q63. Nitrogen fixation by the microorganisms can be detected by adopting the approach of ?

a) Demonstrating growth in a nitrogen free medium

b) Cultivating the microorganisms in the presence of nitrogen labeled with isotropic nitrogen

c) Measuring $^{15}N_2$ by mass spectrometer

d) All of the above

Q64. Which of the following is not the biofertilisers producing bacteria?

a) Nostoc b) Anabaena

c) Both (a) and (b) d) Clostridium

Q65. Which of the following is capable of oxidizing sulfur to sulfates?

a) Thiobacillus thiooxidans b) Desulfotomaculum

c) Rhodospirillum d) Rhodomicrobium

Q66. Which of the following microorganism use H_2S as the electron donor to reduce carbon dioxide?

a) Chromaticum b) Chlorobium

c) Both (a) and (b) d) Rhodomicrobium

Q67. Nitrifying bacteria can not be isolated directly by the usual techniques employed to isolate hetrotrophic bacteria. The reasons may be due to?

a) Slow growth b) Fast growth

c) Medium growth d) None of these

Q68. The population of algae in soil is __________ that of either bacteria or fungi.

a) Generally smaller than b) Generally greater than

c) Equal to d) None

Q69. The physical structure of soil is improved by the accumulation of?

a) Mold mycelium b) Minerals

c) Water d) All of these

Q70. The enzyme nitrogenase consists of?

a) Dinitrogenase b) Dinitrogenase reductase

c) Both (a) and (b) d) None of these

Q71. The crops which are involved in nitrogen fixation are?

a) Legume b) Soybean

c) Bean and lupine d) All of these

Q72. Which of the following fungi on infecting crop roots can improve their uptake of phosphorus and other nutrients?

a) *Saccharomyces cerevisiae* b) *VA Mycorrhiza*

c) *Candida torulopsis* d) *Aspergillus niger*

Q73. The microbial ecosystem of soil includes?

a) Biotic components of soil

b) Abiotic components of soil

c) Biotic and abiotic components of soil

d) None of the above

Q74. Bacteria are likely to be more prevalent in soils of vineyards, orchards and apiaries.

a) True b) False

Q75. Which of the following soil microorganism is involved in the reduction of sulfates to H_2S?

a) *Thiobacillus thiooxidans* b) Desulfotomaculum

c) Rhodospirillum d) Rhodomicrobium

Q76. Microbe i.e *Burkholderia* and *Ralstonia* sps which fix nitrogen in leguminous plants belong to ?

a) Beta- probacteria b) Alpha- proteobacteria

c) Rhizobium group d) Nitrobacteria

Q77. Organisms that obtain energy from the oxidation of inorganic compounds are called?

a) Chemolithotrophs b) Oragnotrophs

c) Lithothrophs d) None of the above

Q78. In 1904 for the first time who used the term rhizosphere?

a) Louis pasteur b) Lister

c) Beijerinck d) L. hiltner

Q79. The phenomenon of loss of organic and inorganic compound from root surface due to unfavorable condition?

a) Root effect b) Root signalaing

c) Both A and B d) Root exudation

Q80. Microbe such as Azotobacter and the Cyanobacteria tichodesium fix nitrogen?

a) Aerobically b) Anaerobiclly

c) Both A and B

Q81. ..... any organism capable of existing as either an autotroph or heterotroph.

a) Mixotrophs b) Autotroph

c) Heterotrophy d) None of these

Q82. Fungi produces which of the following inhibitory toxic product?

a) Cyanide b) Fatty acids

c) Methane d) Sulphides

Q83. Phosphorous and nitrogen ions generally get depeleted in soil because they usually occur as

a) Neutral ions

b) Negatively charged ions

c) Positively charged ions

d) Both positively and negatively charged but disproportionate mixture.

Q84. Which one is an essential mineral, not constituent of any enzyme but stimulates the activity of many enzymes

a) Zn b) Mn

c) K d) Mg

Q85. In soil, the water available for root absorption is

a) Gravitational water b) Capillary water

c) Hygroscopic water d) Combined water.

Q86. Which of the following element plays an important role in biological nitrogen fixation?

a) Copper b) Molybdenum

c) Zinc d) Manganes.

Q87. When water enters in roots due to diffusion, is termed as

a) Osmosis b) Passive absorption

c) Endocytosis d) Active absorption

Q88. Which of the following genera synthesizes Nod factors in order to activate a plant to allow development of an infection thread?

a) Agrobacterium b) Rhizobium

c) Pseudomonas d) Frankia

Q89. Nitrogen fertilizers disrupt ecosystem structure and function by?

a) Promoting heterotrophic growth causing an imbalance in CO_2 levels.

b) Decreasing filamentous fungal development which causes loss of soil crumb structure and subsequent soil fertility.

c) Causing more antibiotic producing bacteria to grow and produce antibiotics which stunt the growth of plants.

d) Reducing the number of nitrogen fixing bacteria in soils.

Q90. Which of the following comes under the category of positive association?

a) Neutralism b) Parasitism

c) Commensalism d) Ammensalism

Q91. True for Humic acid?

a) A dark-brown humic substance that is soluble in water only at higher soil pH values and is of greater molecular weight than fulvic acid. Humic acid may remain for centuries in undisturbed soil

b) A black humic substance that is not soluble in water at any pH, has a high molecular weight, and is never found in base-extracted liquid humic acid products.

c) Both A and B

d) Neither A neither B

Q92. Fluvic acid:

a) A yellow to yellow-brown humic substance that is soluble in water under all pH conditions and is of low molecular weight

b) A yellow to dark-brown humic substance that is insoluble in water under all pH conditions and is of low molecular weight

c) A yellow to yellow-brown humic substance that is soluble in water under acidic conditions and is of high molecular weight

d) A yellow to yellow-brown humic substance that is soluble in water under alkaline conditions and is of low molecular weight

Q93. What is the best pH of the soil for cultivation of plants?

a) 3.4 – 5.4 b) 6.5 – 7.5

c) 4.5 – 8.5 d) 5.5 – 6.5

Q94. Roots of which plant contains a red pigment which have affinity for oxygen

a) Carrot b) Soyabean

c) Mustard d) Radish

Q95. Sulphur is an important nutrient for optimum growth and productivity in

a) Oilseed crops b) Cereals

c) Pulse crops d) Fibre crops

Q96. Lytic enzymes which degrade the cell wall of other microbes is scereted by which of the following microorganism?

a) Fungi b) Algae

c) Staphylococcus d) Myxobacteria

Q97. The degradation of complex molecules in soil by fungi for utilization by bacteria is an example of which type of association?

a) Neutralism b) Mutualism

c) Commensalism d) Antagonism

Q98. The term Mycorrhiza is used by

a) Ruinen b) Hiltner

c) AB Frank d) Winogradsky

Q99. Winogradsky column is used

a) For isolation of purple and green phototrophic bacteria and other anaerobes
b) To measure the BOD
c) For the production of Sulphur
d) For the enumeration of Bacteria

Q100. In which depth in a garden soil does the maximum number of microorganisms per gram occur?

a) At a depth of 3-8 cm.   b) At a depth of 1-3 cm
c) At a depth of 8-12 cm   d) At a depth of 12-25 cm

ANSWERS

1.	d	2.	b	3.	b	4.	d	5.	c
6.	a	7.	d	8.	a	9.	c	10.	c
11.	a	12.	b	13.	b	14.	b	15.	a
16.	c	17.	a	18.	d	19.	c	20.	a
21.	d	22.	a	23.	b	24.	c	25.	d
26.	a	27.	b	28.	a	29.	c	30.	b
31.	d	32.	d	33.	b	34.	a	35.	a
36.	a	37.	b	38.	c	39.	d	40.	b
41.	b	42.	a	43.	b	44.	b	45.	b
46.	c	47.	a	48.	b	49.	d	50.	a
51.	c	52.	a	53.	c	54.	c	55.	d
56.	c	57.	c	58.	d	59.	c	60.	b
61.	b	62.	a	63.	d	64.	d	65.	a
66.	c	67.	a	68.	a	69.	d	70.	c
71.	d	72.	b	73.	c	74.	b	75.	d
76.	a	77.	c	78.	d	79.	d	80.	a
81.	a	82.	a	83.	b	84.	c	85.	b
86.	b	87.	b	88.	b	89.	b	90.	c
91.	a	92.	a	93.	d	94.	a	95.	a
96.	a	97.	c	98.	c	99.	a	100.	d

Unit-II

History of Microbiology

Q1. Father of Microbiology?

a) Robert Koch
b) Louis Pasteur
c) Antony Philips Van Leeuwenhoek
d) A. Jenner

Q2. Golden age of Microbiology?

a) 1856 to 1913
b) 1857 to 1914
c) 1858 to 1915
d) 1859 to 1916

Q3. The first direct demonstration of the role of bacateria causing disease come from the study of Anthrax by?

a) Robert Koch
b) A. Jenner
c) Louis Pasteur
d) Edward jenner

Q4. A porcelain bacterial filter in 1884 was constructed by ?

a) Charles Chamberland
b) Lister
c) Beijerinck
d) Francisco Redi

Q5. Who discovered the existence of heat resistant bacterial endospore ?

a) Louis pasteur
b) leeuwenhoek
c) Rosenbach
d) John tyndall

Q6. Who proved that microbial growth could occur without air contamination?

a) Louis pasteur
b) Felix pouchet
c) lister
d) John Tyndall

Q7. Who suggest that disease was caused by invisible living creatures?

a) Passet

b) Sir Alexander ogston

c) Roman luertis and girolamo fracastaro

d) None of the above

Q8. The comparison of the ribosomal r RNA begun by,

a) Carl woese b) A. gartner

c) Miquel d) John Needham

Q9. In which year the first drawing of a microorganisms was published in Micrographia by Robert hookes?

a) 1665 b) 1666

c) 1667 d) 1668

Q10. Who design first sealing glass flask ?

a) John Needham b) Lazzaro Spallanzani

c) Louis Pasteur d) Robert Koch

Q11. In 1845 who proved that the great potato blight of Ireland was caused by water mold?

a) Marie von ermengem b) S.C. Prescott

c) M. W Beijerinck d) Sergei winogradsky

Q12. Dimitri Ivanowski and Martinus Beijernick used the filter to study which virus?

a) TMV b) Bacteria

c) Yeast d) Both A and C

Q13. Anaerobic nitrogen fixing soil bacteria was first isolated by which one of the following the scientist?

a) M. W Beijerinck b) Sergei winogradsky

c) Lazzaro spallanzani d) Leeuwenhoek

Q14. From the following scientist who first developed the enrichment culture techniques and the use of selective media?

a) Beijernick and winogradsky ?

b) Robert koch

c) Louis pasteur

d) None of the above

Q15. Who was the first person to extensively describe micro-organism?

a) Louis pasteur

b) Antony van leeuwenhoek

c) Robert koch

d) Beijernick and winogradsky

Q16. Koch's posulates and molecular koch's posultes are used to proves a direct relationship B/w a suspected?

a) Pathogen and Enviroment

b) Disease and Enivroment

c) Pathogen and Disease

d) None of the above

Q17. Who showed for the first time that fermentation were caused by microorganisms and some microorganisms could live in the absence of the oxygen?

a) Pasteur b) Lister

c) Robert koch d) Antony van leeuwenhoek

Q18. The role of microbes in C, N, and S cycle was first studied by?

a) Antony van leeuwenhoek

b) Robert koch

c) Winogradsky and Beijerinck

d) Pasteur

Q19. What is the function of the Fts (tublin protein) a prokaryotic protein?

a) Cell shape b) Cell division

c) Cell nutrient d) Cell maintence

Q20. Christian Gram developed the Gram stain in which year?

a) 1883 b) 1882

c) 1884 d) 1885

Q21. The spore often is surrounded by a thin, delicate covering called the

a) Exosporium b) Endosporium

c) Spore coat d) Spore cortex

Q22. A ————————— lies beneath the exosporium

a) Exosporium b) Cortex

c) Spore coat d) Endosporium

Q23. A number of gram positive bacteria can from a special resistant dormant structure called a?

a) Exospore b) Endospore

c) Inclusion bodies d) Exclusion bodies

Q24. What is the size of bacaterial flagella>

a) 0.01 to 0.02 b) 0.02 to 0.04

c) 0.04 to 0.06 d) 0.6 to 0.08

Q25. Poly – beta – hydroxybutyrate (PHb) Which can serve as a reserve carbon and energy source is satined with which dye

a) Gram stain b) Dilute methlene blue

c) Nile blue d) Sudan black

Q26. Endospore which is a dormant structure of Gramm positive bacteria, endospaore contains a large amount of————?

a) Dipicolinc acid b) Calcium

c) Protien d) Amino acid

Q27. Genetic material of Virus is?

a) Only DNA

b) Only RNA

c) Both DNA And RNA

d) Either DNA or Either RNA

Q28. Cauliflower mosaic Virus conatin?

a) DNA

b) RNA

c) BOTH

d) Either DNA or either RNA

Q29. Other name of Blue Green Algae is ?

a) Rhizobium

b) Cyanobacteria

c) Frankia

d) Proteobacteria

Q30. Due to discoverey of which of the following in 1980 the evolution termed as RNA world ?

a) mRNA, tRNA, rRNA Synthesis protien

b) In some virus RNA is genetic material

c) RNA have an enzymatic activity

d) RNA is not found in all cell

Q31. TMV is approximately in size?

a) *~200 nm in length and ~18 nm in diameter.*

b) *~100 nm in length and ~19 nm in diameter.*

c) *~300 nm in length and ~18 nm in diameter.*

d) *~500 nm in length and ~16 nm in diameter.*

Q32. Scientists who proved DNA and not a protein as the genetic material?

a) Hershey and chase

b) Beadle and Tatum

c) Pasteur

d) Jenner

Q33. Reserved food martial in cyanobacteria is ?

a) Protien b) Sugar

c) Amino acid d) Glycogen

Q34. Which one of the following is a viral disease?

a) Red Rot

b) Wilt

c) Stewart's Wilt of Corn

d) Barley Yellow Dwarf of Wheat

Q35. The cyanobacteria of great nutritional value being marketed today is ?

a) *Serytonema* b) Spirogyra

c) Spirulina d) Stigonema

Q36. Protien coat of a virus enclosing nucleic acid is ?

a) Capsid b) Envelope

c) Membrane d) Outer membrane

Q37. Endospore stained by which dye?

a) Malachite green b) Gramm's stained

c) Sudan Black d) Methylne blue

Q38. Mycoplasma are free living organism and small organisms having size approximately?

a) 0.10 to 0.4 μm in diameter

b) 0.15 to 0.3 μm in diameter

c) 0.12 to 0.5 μm in diameter

d) 0.16 to 0.7 μm in diameter

Q39. Which one of the smallest organisms capable of autonomus growth and reproduction?

a) Mycoplasma b) Virus

c) Viroid d) None of the above

Q40. A Little leaf of brinjal is caused by ?

a) Mycoplasma b) Virus

c) Fungus d) Algae

Q41. Cyanophages were discovered by ?

a) Safferman and Morris b) Schroeter and Gessard

c) Hansen d) Klein Berger

Q42. Consider the following regarding the reason for the fact that now the cyanobacteria are kept in Monera not in plantae

1. They are prokaryotic
2. The cell wall of the cyanobacteria has peptidoglycan
3. They can fix the atmoshpheric nitrogen the correct explanation would be

a) 1 and 2 only b) 1 and 3 only

c) 2 and 3 only d) 1,2 and 3 only

Q43. Besides a paddy fileds, cyanobacteria are also found inside of vegetative parts of ?

a) Pinus b) Cycas

c) Equistem d) Psilotum

Q44. Hormogonia are produced by ?

a) *Cyanbacteria* b) *Bacillus* spp.

c) *Nitrobacter* d) *Clostridium* spp.

Q45. Blue Green Algae (BGA) growing on surface are rich in?

a) Phycocyanin b) Phycoerythrin

c) Carotenes d) Xanthophylls

Q46. The amino acid found only in bacteria and blue green algae is ?

a) Methionine b) Diamino-pimelic acid

c) Aspartic acid d) Glutamic acid

Q47. Bacteria multiply mainly by——

a) Transverse binary fission

b) Longitudinal binary fission

c) Oidia

d) Conjugation

Q48. The principle light trapping pigment molecule in plants , algae and cyanobacteria is?

a) Chlorophyll a b) Chlorophyll b

c) Porphyrin d) Rhodapsin

Q49. Antiseptic Method were first introduced by?

a) Lister b) Pastuer

c) Antony d) Griffth

Q50. Endospore are actually help in?

a) Multiplication b) Dispersal and Penetration

c) Variation d) All of these

Q51. The cells in some filamentous cyanobacteria that are specialized for nitrogen fixation are called?

a) Phycobillisomes b) Chromatophore

c) Grana d) Heterocyst

Q52. Phagocytosis phenomenon was discovered by?

a) Rober koch b) Lazzaro spllanzi

c) Jenner d) Élie Metchnikoff

Q53. Hanging drop method for motility study was first introduced by?

a) Leeuwenhock b) Robert koch

c) Jenner d) Pasteur

Q54. The bacterium most commonly used in genetic engineering is

a) *E. coli* b) *Pseudomonas*

c) *Bacillus* d) *Yeast*

Q55. Endotoxin produced by Gram negative bacteria is present in?

a) Flagella b) Cytoplasm

c) Plasma membrane d) Lipid polysaccharides

Q56. The viruses that live as parasites on bacteria are?

a) Virion b) Bacteriophage

c) Viroid d) Prion

Q57. The family Chlorobiaceae contains ?

a) Purple sulfur bacteria b) Green sulfur bacteria

c) Both d) None of these

Q58. Archea stain is a —

a) Gram positive

b) Gram negative

c) Either Gram positive or either Gram Negative

d) None of these

Q59. Type strain is used for referring to?

a) Species b) Genus

c) Family d) Division

Q60. Archaea are in nature

a) Aerobic b) Anaerobic facultative

c) Strict anaerobic d) all of them

Q61. The Achaea differ from both bacteria and eukaryote in having

a) Branched chain hydrocarabon attached to glycerol by ether linkage

b) Branched chain hydrocarabon attached to glycerol by ester linkage

c) Branched chain hydrocarabon attached to fatty acid by ether linkage

d) Branched chain hydrocarabon attached to fatty acid by ester linkage

Q62. Pentacyclic rings are used by thermophillics archaea to help maintain then delicates liquid crystalline balance of to membrane at

a) High temperature b) High saline

c) High alkaline d) High acidic condition

Q63. The genome size of *Bacillus subtilis* is —

a) 120 million bp b) 4.20 million bp

c) 300 million bp d) 250 million bp

Q64. The purpose of the swan neck flask that **Louis Pasteur** designed to disprove spontaneous generation is to?

a) Allow the multiplication of microbes in the broth

b) Implicate the role of flies in the development of maggots on rotting meat

c) Pasteurize the meat broth

d) Trap the microbes and prevents them from reaching the broth

Q65. The Archaeal do not appear to degrade glucose by the pathway of which

a) Embden Meyerhof pathway

b) Glycosis

c) Entner- Doudoroff pathway

d) None of the above

Q66. Which enzyme do not present in archaea?

a) Pyruvate dehydrogenase b) Pyruvate oxidoreductase

c) Acotinase d) 6-phosphofrcutokinase

Q67. Methanogenic bacteria are –

a) Strict anaerobes b) Strict aerobes

c) Facultative aerobes d) Facultative aerobes

Q68. Thermoplasma genera of archeal taxonomy is —

a) Strict anaerobic b) Facultative anaerobic

c) Strict aerobes d) Facultative aerobics

Q69. The Halobacteria the extreme halophiles or the most obvious distinguish trait of this family its absoltute dependence on a high conc of NaCl, these prokaroytes require at least

a) 1.5 m Nacl b) 2.5 M Nacl

c) 1 M Nacl d) 2 M Nacl

Q70. Which one of the following fungi forms chlamydospore type asexual spores?

a) *Aspergillus* b) *Fusarium*

c) *Puccinia* d) *Albuga*

Q71. On the basis of their characteristics, which one of the following method refers to grouped microorganism by numerical methods of taxonomics units into taxa?

a) Genotypic b) Numerical taxonomy

c) Phonctic taxonomy d) Phylogenetic taxonomy

Q72. Protozoa have no charactcteristics like one of the following?

a) Frequently possess cilia or flagella

b) Most exhibit asexual reproduction

c) Most similar to plants

d) Singled celled eukryotic organisms

Q73. Regarding the fungi, except which one of the following is true?

a) Fungi are eukaryotes b) Fungi are photosynthetic

c) Molds form hyphae d) Yeast are unicellular

Q74. *E. coli* and *Salmonella* belongs are the most commonly found pathogens of animals. They belongs to which one of the following class?

a) α-proteobacteria b) β-proteobacteria

c) γ-proteobacteria d) δ-proteobacteria

Q75. Mycoplasma is bacterium without cell wall having cell memebfrane on outerside. This bacterium belongs to which one of the following class?

a) Bacilli b) Clostridia

c) Mollicute d) Proteobacteria

Q76. Largest bacterial genera in domain of bacteria are belongs to which one of the following ?

a) *Actinobacteria* b) *Bacilli*

c) *Mollicutes* d) *Clostridia*

Q77. Which one of the following is a group of microbes that form sulfur granules inside their cell?

a) Cyanobacteria b) Greensulfur bacteria

c) Purple non sulfur bacteria d) Purple sulfur bacteria

Q78. Which one of the following molecule has the highest absorption maxima at 260 nm wavelength?

a) dsDNA b) dsRNA

c) ssDNA d) ss RNA

Q79. Which of the following is true about alpha proteobacteria except?

a) Ammonia oxidation b) Gram positive

c) Methylotrophy d) Nitrogen fixation

Q80. Root nodule are formed in leguminous plants by which one of the following microorganism?

a) *Agrobacterium* b) *Nitrobacter*

c) *Rhizobium* d) *Rhodospirilla*

Q81. Which one of the following is known to gain energy by heterolactic fermentation ?

a) *Bacillus* b) *Lactobacillus*

c) *Leuconostoc* d) *Staphlyococcus*

Q82. Which one of the microorganism belongs to low G+C Gram+ve bacterial group?

a) *Actinobacter* b) *Salmonella*

c) *Bacillus* d) *Anabaena*

Q83. Fluorescent Pseudomonas belongs to which one of the following group?

a) Alpha-proteobacteria b) beta-proteobacteria

c) gamma-proteobacteria d) delta-proteobacteria

Q84. Which one of the following fungi is an example of Oospores types sexual spores?

a) *Phytopthora* b) *Pythium*

c) *Albugo* d) Both a and b

Q85. Which one of the following microbes is known to gain energy by substrate level phosphorylation?

a) *Bacillus* b) *Lactobacillus*

c) *Satphlylococcus* d) *Clostridia*

Q86. Which one of the following is not found in mycoplasma that distinguished with prokaryotes?

a) Cell-wall b) Chitin

c) Murrain d) Protein

Q87. Which one of the following is not a function of a cell membrane in a bacterial cell?

a) Nutrient uptake b) Protein secretion

c) Protein synthesis d) Waste excretion

Q88. Which one the following amino acid is found in cell wall of bacteria, except?

a) D-alanine

b) D-gluatmic acid

c) L-alanine

d) Mesodiamnino pimpelic acid

Q89. Which one of the following pathway is found in most plant and animal pathogenic bacteria ?

a) Type-I b) Type-II

c) Type-III d) Type-IV

Q90. Which one of the following layers of material presents outside of the bacterial cell and can't be easily washed off?

a) Capsule b) Glycocalyx

c) Sheath Layer d) Slime Layer

Q91. Which one of the following molecule is least likely to attach with polar lipids of archae?

a) Methyl b) Phosphate

c) Sugar d) Sulfate

Q92. For degradation of different metabolites in natural habitat, which of the following plasmid helps bacteria?

a) Resistant plasmid b) Metablic plasmid

c) Natural plasmid d) None of the above

Q93. Ester linked membrane lipids presents in which one of following?

a) Archaea b) Bacteria

c) Eukarya d) Both B and C

Q94. Which one of the following is not presented in chemical composition Gram negative cell wall?

a) Glycoprotien b) Lipopolysaccharides

c) Lipoprotien d) Peptidoglycan

Q95. Which statement is correct about the Thermoactinomycetes a filamentous bacteria ?

a) grow at 40-50 degree and can decompose the cellulose

b) the spores are endospore and survive 95 degree for 10 mins, are highly refractile, and conatins dipicolinc acid

c) Form compacted group of erect hyphae

d) Strikingly different from all other filamentaous bacteria

Q96. The word mycorrhiza is derived from the Greek word, meaning ———?

a) Short root b) Bacterial root

c) Fungus root d) Infected root

Q97. Gram negative nonphototrophic bacteria form an external sheath that covers the chains or trichomes are?

a) Fruiting bacteria b) Archaeobacteria

c) Sheathed bacteria d) Gliding bacteria

Q98. All of the following scientist got Nobel Prize for their contribution in the field of Microbiology except?

a) Antony Van Leeuwenhoek

b) Elie Metchnikoff

c) Paul Ehrlich

d) Robert Koch

Q99. Purple phototrophic bacteria conatins bacteriochlotophyll?

a) Type b b) Type a or b

c) Either a or either b d) None of the above

Q100. Whichone of the following is best studied conjugative plasmid?

a) F plasmid b) P plasmid

c) C plasmid d) R plasmid

Q101. In wich one of the following 80S ribosomes sediments value present?

a) Archaea b) Bacteria

c) Eukarya d) both a and b

Q102. Pure culture of bacteria were first obtained by ———— serial dilution in liquid media?

a) Cienkowski b) Joshep Lister

c) Muntz d) None of the above

Q103. The purpose of swan-necked flasks that Louis Pasteur designed to disprove spontaneous generation is to ?

a) Allow the multiplication of microbes in the broth

b) Implicate the role of flies in the development of maggots on rotting meat

c) Pasteurize the meat broth

d) Trap the microbes and prevent them from reaching the broth

Q104. The event that triggered the development and establishment of microbiology as a science is the —

a) Development of Microscope

b) Germ theroy of disease

c) Spontaneous generation

d) Use of disinfectants

Q105. The epidemic that infected Europe, Middle East and North Africa and killed tens of millions of people was known as the Black Death.

The disease was caused by?

a) Anthrax b) Breathing of foul air

c) Small pox d) Bubonic plague

Q106. Scientist who demonstrate that open tubes of broth remained free of microbe when air was free of dust?

a) Tyndal b) Lousi Pasteur

c) Robert Koch d) Schroeter and Gessard

Q107. L-forms are discovered by —

a) Antony von leeuwenhock b) Kelinberger

c) Louis Pasteur d) Robert Koch

Q108. The concept of spontaneous generation related to which one of the following theory ?

a) Theory of abiogenesis b) Theory of anagenesis

c) Theory of biogenesis d) None of the above

Q109. Which of thc following scientist discovered the viral infection of bacteria ?

a) Beijernick b) De; herelle

c) F.W. Twort d) Dmitri Ivanosky

Q110. The structure of the molds was described in 1664 by which one of the following scientist?

a) Robert Hooke b) Leonardo da vinci

c) leeuwenhoek d) None of the above

Q111. Who among the following first isolated genomics DNA from the bacterial cell?

a) Freidrick Sanger b) Friedrich Miescher

c) James W Watson d) Thomas czeck

Q112. Which one of the following layers of material presents outside of bacterial cell and can't be easily washed off?

a) Capsule b) Glycocalyx

c) Shaeath layer d) Slime layer

ANSWERS

1.	c	2.	b	3.	c	4.	a	5.	d
6.	d	7.	c	8.	a	9.	a	10.	b
11.	c	12.	a	13.	a	14.	a	15.	b
16.	c	17.	a	18.	c	19.	a	20.	c
21.	c	22.	a	23.	b	24.	b	25.	d
26.	a	27.	d	28.	a	29.	b	30.	c
31.	c	32.	a	33.	d	34.	d	35.	c

36.	a	37.	a	38.	b	39.	a	40.	a
41.	a	42.	a	43.	b	44.	a	45.	a
46.	b	47.	a	48.	a	49.	a	50.	b
51.	d	52.	d	53.	a	54.	a	55.	d
56.	b	57.	b	58.	c	59.	a	60.	d
61.	d	62.	a	63.	b	64.	d	65.	a
66.	a	67.	a	68.	b	69.	d	70.	b
71.	b	72.	b	73.	a	74.	b	75.	c
76.	b	77.	b	78.	d	79.	a	80.	c
81.	c	82.	c	83.	c	84.	d	85.	b
86.	a	87.	b	88.	c	89.	d	90.	b
91.	a	92.	b	93.	d	94.	c	95.	a
96.	c	97.	c	98.	a	99.	b	100.	a
101.	c	102.	b	103.	d	104.	a	105.	d
106.	a	107.	b	108.	a	109.	c	110.	a
111.	b	112.	b						

Unit-III

Microbial Ecology and Physiology

Q1. The place where microorganism actually lives and metabolically active is called

a) Microenvironment b) Macroenvironment

c) Niche d) Habitat

Q2. In *Pseudomonas aeruginosa* the major signaling molecule in biofilm is ——————

a) Homoserine lactone b) Acyl homeserine

c) Na d) K

Q3. Which of these is not an advantage of biofilm formation ——————

a) Defence

b) Allow cells to remain in a favourable niche

c) Allow bacteria to live in a close association

d) All of these

Q4. Furamones are ——————

a) Biofilm Preventive b) Antocoagulant

c) Biofim Promoter d) All of these

Q5. Major factor that affect microbial activity in soil is called ——————

a) pH b) Temperature

c) Water d) All of these

Q6. Algae floating freely in water are called ————

a) Phytoplankton b) Zooplankton

c) Benthic Algae d) All of these

Q7. Algae attached to bottom are called ————

a) Benthic algae b) Floating algae

c) Benzoic algae d) All of these

Q8. In lakes the upper layer of water is called ————

a) Epilimnion b) Thermocline

c) Hypolimnion d) All

Q9. Organisms that tolerate high pressure are called ————

a) High pressure tolerant b) Barotolerant

c) Barophilic d) None of these

Q10. High pressure tolerant microbes are called ————

a) Barophilic b) Basophilic

c) High pressure tolerant d) All of the above

Q11. Hydrothermal vents are ————

a) Hot spring vents b) High pressure vents

c) Underwater hot springs d) None of these

Q12. Vents which emit hydrothermal fluid at a temp. 270-380 c are called ————

a) Hot vents b) Black smoke

c) Both a& b d) All of the above

Q13. Methanogens are

a) Strict anaerobes b) Strict aerobes

c) Obligate anaerobe d) Obligate aerobe

Q14. Process in which two or more microorganisms cooperate in the degradation of some compounds and focused on the energetic behind the process is called

a) Symtrophy b) Syntrophy

c) Asymtrophy d) All of these

Q15. Within the cells of protozoa *Trichomonas* which type of microorganisms present ——————

a) Methanogens b) Barophiics

c) Acidophilic d) None of these

Q16. The conversion of nitrate to gaseous nitrogen compound is called ——————

a) Denitrification b) Nitrification

c) Both d) None of these

Q17. If protein rich material like sewage, manure are added in soils, rate of nitrification will

a) Increase

b) Decrease

c) Remain constant

d) Sometime will increase and some time decrease

Q18. Ammonia can be catabolized anaerobically by a process called ——————

a) denitrification b) nitrification

c) anammox d) none

Q19. Common form of iron in nature is ——————

a) pyrite b) bauxite

c) both d) none of these

Q20. Root surface is called ——————

a) Rhizosphere b) Rhizoplane

c) Phylosphere d) None of these

Q21. Bacterial count is always higher in——————

a) Rhizoplane b) Rhizosphere

c) Phyloplane d) Phylosphere

Q22. Lichens consist of ——————

a) Bacteria and fungus b) Fungus and algae

c) Bacteria and virus d) Bacteria and algae

Q23. In lichens producer is ——————

a) Algae b) Fungi

c) Bacteria d) None of these

Q24. Mycorrhiza refers to association between ——————

a) Plant roots and fungi

b) Association between algae and fungi

c) Association between plant root and fungi

d) None of these

Q25. Fungus cells form an extensive sheath around the outside surface called ——————

a) Ectomycorrhiza b) Endomycorrhiza

c) Lichens d) All of the above

Q26. Rhizobium, bradyrhizobium, sinorhizobium, mesorhizobium, azorhizobium and proteobacteria are ——————

a) Gram negative nodule forming bacteria

b) Gram positive nodule forming

c) Gram negative non-nodule forming

d) Gram positive non- nodule forming

Q27. In nodules precise oxygen levels are maintained by oxygen binding proteins ——————

a) Leghemoglobin b) Oxyhemoglobin

c) Legoxyhemoglobin d) None of these

Q28. *Bradyrhizobium japonicum* nodulates which plant ——————

a) Pea b) Bean

c) Soyabean d) Alfalfa

Q29. Swollen, misshapen and branched forms develops when rhizobia multiply within plants are called ——————

a) Nodule b) Bacteroids

c) Galls d) Crowns

Q30 .... Genes direct the steps in nodulation of a legume by a specific rhizobia strains are called

a) Nif genes b) Nod genes

c) Oxyhaemoglobin d) Leghaemoglobin

Q31. NOD factors are involved in ___________

a) Root hair curling and nodule formation

b) Nodulation

c) Root hair curling

d) All of the above

Q32. First stable product of nitrogen fixation is ___________

a) Nitrogen b) Ammonia

c) Nitrate d) Nitrite

Q33. Stem nodulating rhizobia produce ___________

a) Bacteriochlorophyll b b) Bacteriochlorophyll a

c) Both a & b d) None of these

Q34. Water fern azolla contain N2 fixing cyanobacteria ___________

a) *Anabaena azollae* b) Nostoc

c) Both d) None of these

Q35. Anus plant contain nitrogen fixing organism

a) *Rhizobium* b) *Frankia*

c) *Anabaena* d) *Azollae*

Q36. Nitrogenase enzyme is prevented by oxygen damage through _____

a) Heterocyst b) Sheath

c) Endospore d) None of these

Q37. Non leguminious nitrogen fixing symbionants are ———

a) *Rhizobium* and *Bradyrhizobium*

b) *Azolla anabaena* and *Frankia*

c) *Nostoc*

d) *Anabaena nostoc*

Q38. Hairy root disease is caused by ———

a) Tumefaciens b) Rhizogenes

c) Both a & b d) None of these

Q39. Ri plasmid is necessary for the induction of ———

a) Hairy root b) Crown gall

c) Tumer formation d) None of these

Q40. Which of these is an obligatory relationship ———

a) Cooperation b) Commensalism

c) Mutualism d) All

Q41. Relationship in which one microorganism is benefited while the other is neither helped nor harmed is ———

a) Symbiosis b) Commensalism

c) Cooperation d) None of these

Q42. Relationship in which both microorganisms are benefited but which is non obligatory ———

a) Cooperation b) Symbiosis

c) Predation d) Commensalism

Q43. Relationship in which one pair is benefited from the other while the other and the host is usually harmed is called ———

a) Predation b) Parasitism

c) Amensalism d) Commensalism

Q44. Production of antibiotics that kill or inhibit suceptible microorganism is an example of ———

a) Amensalism b) Commensalism

c) Symbiosis d) Cooperation

Q45. Nitrification is an example of————

a) Commensalism b) Symbiosis

c) Cooperation d) Amensalism

Q46. Protozoa-termite relationship is an example of————

a) Amensalism b) Mutualism

c) Commensalism d) All of the above

Q47. Following diagram indicates————

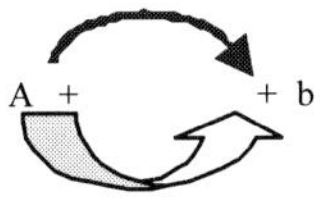

A + + b

a) Parasitism b) Mutualism

c) Competetion d) Predation

Q48. Nutrient rich lakes are called————

a) Oligotrophic b) Eutrophic

c) Oligoeutrphic d) None of these

Q49. Epilimnion is————

a) Aerobic b) Anaerobic

c) Obligate aerobic d) None of these

Q50. Epilimnion and hypolimnion are separated by a zone of transition called————

a) Thermocline b) Thermostat

c) Thermocele d) None of these

Q51. In which growth stage balanced growth occur————

a) Exponential b) Lag

c) Death d) All of the above

Q52. Population doubling time is called————

a) Generation time b) Growth time

c) Both d) None of these

Q53. Formula for calculating mean growth rate, k is ————

a) K=n/t b) G=i/k

c) Nt=2n0 d) None of these

Q54. Antibiotic production occurs in which growth phase ————

a) Lag b) Log

c) Exponential d) Death

Q55. Microbial growth curve is........ in shape ————

a) Sigmoidal b) Semi-sigmoidal

c) Oval d) All of these

Q56. Which of the following is not a technique for the measurement of cell mass

a) Coulter counter

b) Hemocytometer

c) Cell turbidity measurement

d) Petroff-Hauser counting

Q57. Large microorganisms such as protozoa, algae are counted by ————

a) Coulter counter b) Petroff hauser counting

c) Spectrophotometer d) All of these

Q58. Chemostat is a type of ————

a) Countinous culture b) Discountinous culture

c) Both a & b d) None of these

Q59. If the flow rate of vessel is 30(ml/hr) and volume is 100 ml then what will be the dilution rate ————

a) 1.1 hr-1 b) 0.3 hr

c) 0.3 hr-1 d) None of these

Q60. Open system of cultivation is called ——————

a) Fed batch culture b) Batch culture

c) Countinous culture d) All of the these

Q61. Batch culture is also known as ——————

a) Closed system b) Open system

c) Semi-closed system d) All of these

Q62. Which is semi-closed system ——————

a) Fed-batch b) Batch

c) Countinous d) All of these

Q63. In which type of culture system cell density remain constant ——————

a) Constant b) Batch

c) Fed batch d) None of these

Q64. In which type of culture vessel photoelectric turbidostat is used to monitor cell density ——————

a) Turbidostat b) Chemostat

c) Both a&b d) All of these

Q65. Two factors work in chemostat ——————

a) Dilution rate & limiting nutrient concentration

b) Temperature

c) pH

d) All of these

Q66. In which type of culture bacterial growth can be maintained at exponential growth phase for long time ——————

a) Fed batch b) Batch

c) Countinous d) All of these

Q67. Diphasic growth is intervened by a short ——————

a) Stationary phase b) Lag phase

c) Log phase d) Death phase

Q68. The growth phase during which all bacterial cells of a population are identical and in same stage of cell division is called ——————

a) Synchronous growth b) Diauxic growth
c) Both a & b d) None of these

Q69. How many ATP does one NADH makes during ETC ——————

a) 6 b) 3
c) 2 d) 8

Q70. Role of oxygen in aerobic respiration is ——————

a) E donor b) E acceptor
c) Both d) None of these

Q71. Location of ETC in prokaryotes is ——————

a) Cytoplasm b) Cell membrane
c) Nucleus d) None of these

Q72. Glycolysis occurs in ——————

a) Cytoplasm b) Nucleus
c) Cell membrane d) All of these

Q73. Which of the following generates majority of ATP in cellular respiration

a) ETC b) Glycolysis
c) Krebs cycle d) Oxylate cycle

Q74. Process of conversion of pyruvate to CO2 and OH and 2 ATP molecule is called

a) Alcohol fermentation b) Anaerobic fermentation
c) Fermentation d) Aerobic fermentation

Q75. Process of removal of carboxyl group from a chemical compound is called

a) Deamination b) Decarboxylation
c) Oxylation d) None of these

Q76. Organisms live on organic compounds produced by other microorganisms are known as

a) Autotrophs b) Heterotrophs

c) Organotrophs d) Phototrophs

Q77. NAD + is an ————

a) Electron carrier b) Currency of cell

c) Power house of cell d) None of these

Q78. In which type of respiration electron acceptor is inorganic molecule ————

a) Anaerobic b) Aerobic

c) Both d) None of these

Q79. Glycolysis converts glucose to ————

a) Acetic acid b) Oxaloacetic acid

c) Pyruvic acid d) Pyruvate

Q80. Krebs cycle occur in ————

a) Matrix of Mitochondria b) Cytoplasm

c) Nucleus d) Cell membrane

Q81. Production of ATP using energy derived from the redox reaction of ETC is ————

a) Substaratate level phosporylation

b) Oxidative phosphorylation

c) Both a and b

d) None of these

Q82. Formation of ATP by directly transferring a phosphate group to ADP from an intermediate substrate is called ————

a) Substrate level phosphorylation

b) Oxidative phosphorylation

c) Both a & b

d) None of these

Q83. Organisms that use organic carbon as their carbon source are called ——————

a) Photoautotrophs b) Photoheterotrophs

c) Heterotrophs d) None of these

Q84. Chlorophyll contains ——————

a) Iron atom b) Carbon atom

c) Magnesium atom d) None of these

Q85. Formula for chlorophyll a is ——————

a) $C_{55}H_{72}O_5N_4$ mg b) $C_{50}H_{76}O_5N_4$ mg

c) $C_{55}H_{72}O_5N_1$ mg d) $C_{55}H_{72}O_5N_3$ mg

Q86. Formulla for chlorophyll b is ——————

a) $C_{55}H_{71}MgN_4O_6$ b) $C_{55}H_{70}MgN_4O_6$

c) $C_{55}H_{70}MgN_2O_6$ d) $C_{55}H_{70}MgN_4O_5$

Q87. β-carotenoid formula is

a) C40H56 b) C48H56

c) C40H50 d) C40H54

Q88. Carotenoid absorb light in the ——————

a) Blue region b) Red region

c) Green region d) All of these

Q89 ... chloroplasts contain phycobilioproteins ——————

a) Cyanobacteria and red algae

b) Purple bacteria

c) Green bacteria

d) None of these

Q90. In photosystem 1 light is absorbed by ——————

a) p700 b) p680

c) p600 d) p550

Q91. Oxygenic and anoxygenic reactions occur in ————

a) Membranes b) Chloroplasts

c) Nucleus d) Cell membranes

Q92. In oxygenic photosynthesis path of electron resembles letter ————

a) A b) B

c) Z d) E

Q93. Photosystem 1 and 2 work together in ————

a) Oxygenic photosynthesis

b) Anoxygenic photosynthesis

c) Glyoxalate cucle

d) None of these

Q94. Two key enzymes of Calvin cycle are ————

a) Ribulose biphosphate carboxylase and phosphoribulokinase

b) Ribulose biphosphate carboxylase and phosphatase

c) Phosphoribulokinase and disodium phosphate

d) Phoshoenolpyruvate and ribulose

Q95. Reverse citric acid cycle occur in ————

a) Chlorobium b) Azolla

c) Anabaena d) Nostoc

Q96. Organisms that obtain energy from the oxidation of inorganic compounds are called ————

a) Chemolithotrophs b) Organotrophs

c) Phototrophs d) None of these

Q97. Methanogens are ————

a) Aerobic b) Anaerobic

c) Obligate aerobe d) Stringent anaeorobe

Q98. Synthesis of ATP during photosynthetic electron flow is called ——————

a) Photophosphorylation

b) Oxidative photophosphorylation

c) Non oxidative phosphorylatiion

d) None of these

Q99. During anoxygenic photosynthesis light is absorbed by——————

a) P-870 b) P-440

c) P-670 d) P-879

Q100. When a bacterial cell and mitochondria are treated with cyanide and carbon monoxide what happens initially?

a) Respiration inhibits b) Photosynthesis inhibits

c) Protein synthesis inhibits d) No effect occurs

Q101. The most important energy-yielding reaction for an aerobic organism is

a) Glycolysis b) EMP

c) KDPG d) Both b and c

Q102. Important class of respiratory enzymes ——————

a) NAD b) Cytochromes

c) ATPase d) Hydrolases

Q103. Most bacteria require vitamins as ——————

a) Growth Factors

b) Sources of energy

c) Sources of carbon

d) Sources of electron donars

Q104. Which of these is a trace element for bacteria ————?

a) Mg+2 b) Na+

c) Ca+2 d) Mn+2

Q105. The main product of glycolysis under aerobic conditions is ____________

a) Pyruvate
b) Lactate
c) None of these
d) Both a and b

Q106. The protein moiety of an enzyme is known as

a) Holo enzyme
b) Apo enzyme
c) Co enzyme
d) Enzyme

Q107. Nitrites are oxidized to nitrates by a microorganism ————————

a) Nitrosomonas
b) Nitrosococcus
c) Nitrobacter
d) Azatobacter

Q108. Which of the following expression is correct:

a) K=log (a/a-x) x t_1 - temperature effect
b) K=C_n t-watson's expression
c) K_1/K_2=q(t_2-t_1)-concentration of bactericide
d) X_2 = 4d t i_n (m_o/m)-film coefficient

Q109. An example of competitive inhibition of an enzyme is the inhibition of

a) Succinic dehydrogenase by malonic acid
b) Cytochrome oxidase by cyanide
c) Hexokinase by glucose-6-phosphate
d) Carbonic anhydrase by carbon dioxide

ANSWERS

1.	a	2.	a	3.	d	4.	a	5.	c
6.	a	7.	a	8.	c	9.	b	10.	a
11.	c	12.	c	13.	a&c	14.	b	15.	a
16.	a	17.	a	18.	c	19.	a	20.	b
21.	b	22.	b	23.	a	24.	a&c	25.	a

26.	a	27.	a	28.	c	29.	b	30.	b
31.	a	32.	b	33.	a	34.	a	35.	b
36.	a	37.	b	38.	b	39.	a	40.	c
41.	b	42.	a	43.	b	44.	a	45.	a
46.	b	47.	b	48.	b	49.	a	50.	a
51.	a	52.	a	53.	a	54.	c	55.	a
56.	c	57.	a	58.	a	59.	c	60.	c
61.	a	62.	a	63.	a	64.	a	65.	a
66.	c	67.	b	68.	a	69.	b	70.	b
71.	b	72.	a	73.	a	74.	a	75.	a
76.	b	77.	a	78.	a	79.	c&d	80.	a
81.	a	82.	a	83.	b&c	84.	c	85.	a
86.	b	87.	a	88.	a	89.	a	90.	a
91.	a	92.	c	93.	a	94.	a	95.	a
96.	a	97.	b	98.	a	99.	a	100.	a
101.	d	102.	b	103.	a	104.	b	105.	a
106.	b	107.	b	108.	a	109.	a		

Unit-IV

Environmental Microbiology and Basic Microbiology Techniques

Q1. Which of the following is a neutral stain?

a) Picric acid b) Giemsa

c) Neutral red d) Malachite green

Q2. Peptone water medium is an example for

a) Synthetic medium b) Semisynthetic medium

c) Differential medium d) None of these

Q3. Separation of a single bacterial colony is called

a) Isolation b) Separation

c) None of these d) All of thcsc

Q4. Which of the following induces dimerisation of thymine?

a) X-rays b) U.V. rays

c) γ-rays d) None of these

Q5. *Treponema pallidum* can be best indentified using

a) Fluorescence microscope b) Bright field microscope

c) Dark field microscope d) None of these

Q6. Algae are rich in

a) Carbohydrates b) Proteins

c) Vitamins d) All of these

Q7. L-Lysine is produced from

a) Corynebacterium glutamicum

b) Clostridium botulinum

c) Mycobacterium sps

d) Pseudomonas

Q8. Tyndallisation was proposed by

a) Tyndall b) Pasteur

c) Koch d) Jenner

Q9. Viruses can be cultivated in

a) Lab media b) Broth

c) Living cells d) None of these

Q10. A common laboratory method of cultivating anaerobic micro-organisms is

a) Gas pack system

b) Brewer jar system

c) Pyrogallic acid over the cotton

d) None of these

Q11. *Shigella* was first isolated by

a) Shiga b) Schmitz

c) Sonnei d) Robert Koch

Q12. *Clostridium welchii* is positive for

a) Elek's gel precipitation test

b) Nagler's test

c) Weil felix test

d) Bacitracin test

Q13. Nagler's reaction detects

a) Coagulase b) Hyaluronidase

c) Lecithinase d) None of these

Q14. Salt agar is used for

a) Streptococcus b) Staphylococcus

c) Vibrio d) Shigella

Q15. *D.pneumoniae* can be identified by

a) Microscopic exam b) Culture of sputum/blood

c) Animal inoculation d) All of these

e) None of these

Q16. The diagnosis of tuberculosis is carried out by

a) Emulator
b) Antiformin method
c) Petroff's method
d) Concentration method
e) All of these

Q17. The size of the virus can be determined by

a) Micrography
b) Ultra-centrifugation at high speed
c) Ultra-filteration
d) All of these

Q18. Electron microscope studies does not help in identifying the section of bacterial spore

a) Core
b) Spore cortex
c) Capsule
d) All of these

Q19. Wilson and Blair bismuth sulphite medium is used for the growth

a) *Salmonella typhi*
b) *Shigella dysenteriae*
c) *Vibrio cholerae*
d) *E. coli*

Q20. Electron microscope gives magnification upto ——

a) 100 X
b) 2000 X
c) 50,000 X
d) 2,00,000 X

Q21. Viral infection of bacteria was discovered by

a) De'Herelle
b) F.W. Twort
c) Beijernick
d) Iwanowski

Q22. Eye cannot resolve any image less than

a) 1μm
b) 2μm
c) 7μm
d) 5μm

Q23. Compound Microscope was discovered by

a) A.V. Lewenhoek
b) Pasteur
c) Janssen and Hans
d) None of these

Q24. Electron Microscope was discovered by

a) Prof. Fritz b) Janssen and Hans

c) Knoll and Ruska d) None of these

Q25. Magnification range of light microscope is

a) 1000x – 5000x b) 1000x – 2000x

c) 500x – 1000x d) None of these

Q26. Condensation of light in light Microscope is by

a) Objective b) Condensor

c) Ocular d) All of these

Q27. Light gathering capacity of Microscope is called

a) Numerical aperture b) Angular aperture

c) Both a and b d) None of these

Q28. If 10x and 40x objectives are used (air is the medium), the numerical aperture is

a) 1.5 b) 2.0

c) 1.0 d) 1.8

Q29. The ability of Microscope to distinguish two objects into two separate objects, is called.

a) Resolving power b) Wave length

c) N.A. d) None of these

Q30. Limit of resolution of compound microscope is

a) 0.018 Ao b) 0.1 mm

c) 5 μm d) 1 mm

Q31. Source of light in fluorescence microscopy is from

a) Mercury lamp b) Sunlight

c) Both a and b d) None of these

Q32. The magnification power of electron microscope developed by Knell and Ruska is

a) 10,000x b) 12,000x

c) 15,000x d) 20,000x

Q33. In electron microscope source of electrons is from

a) Mercury lamp b) Tungsten metal

c) Both a and b d) None of these

Q34. The electron passed out from the specimen are called

a) Primary electrons b) Secondary electrons

c) Tertiary electrons d) None of these

Q35. Fluroscent substance used in fluorescent microscopy are

a) Quinine sulphate b) Auramine

c) All of these d) None of these

Q36. Steam Sterilization is carried out with

a) Autoclave b) Hot air oven

c) Radiation chamber d) All od these

Q37. Pasteurization was discovered with

a) Louis Pasteur b) Anotny leeuwenhhok

c) Hugo De varies d) R.Koch

Q38. Flash Pasteurization or high temp. short time pasteurization is carried out at what temp. and time

a) 72°C 15 sec b) 72°C 20 sec

c) 72°C 24 sec d) 70°C 15 sec

Q39. Process of heating milk or milk products at 140-150 c for 1-3 sec is called

a) Flash Pasteurization

b) ultra high temperature sterilization (UHT)

c) Pasteurization

d) Milk sterilaztion

Q40. Dry Heat kills cells by

a) disruption of cell wall

b) Alteration in metabolic profiles

c) oxidation of cell constituents and denaturation of proteins

d) disturbance in protein machinery

Q41. What conditions are used for dry heat sterilization

a) 160-170 c for 2-3 hours

b) 120-150 c for 4 hours

c) 160 c for 5 hours

d) 120c for 2 hours

Q42. Heat Senstive materials are sterilized by

a) Autoclave b) Gamma radiation

c) Filtration d) Dry heat sterilization

Q43. Laminar Air flow have............... filters

a) 0.2 um membrane filters

b) HEPA filters

c) Asbestos filters

d) 0.005um membrane filters.

Q44. Size of particles removed by Hepa filters-

a) 0.003 um b) 0.3 um

c) 0.03 um d) 0.8 um

Q45. In cold sterilization which radiation is used

a) X ray b) Gamma

c) u v d) All

Q46 .... was the first disinfectant used

a) Lysol b) Phenol

c) Ethanol d) Methanol

Q47. Lysol is a

a) Disinfectant b) Antibiotic

c) Both d) None

Q48. Alcohols works by

a) Denaturing proteins and dissolving membrane lipids

b) Disrupting cell walls

c) Disturbing protein machinery

d) None of these

Q49. In the liquid form betapropiolactone is used to sterilize

a) Vaccine & sera
b) Antibiotics
c) Syringe
d) Plastic ware

Q50. Copper sulfate is an effective —

a) Algicide
b) Fungicide
c) Bactericidal
d) Viricidal

Q51. Which gas is used for the sterilization of heat sensitive materials like syringes, plastic dishes etc

a) Ethylene oxide
b) Hydrogen peroxide.
c) Ozone
d) Ammonia

Q52. Best method of getting pure culture is

a) Streak plate
b) Pour plate
c) Spread plate
d) All

Q53. Which aladehyde is most powerful disinfectant

a) Formaldehyde and glutaraldehyde
b) Acetaldehyde and formaldehyde
c) Both of these
d) None of these

Q54. For the transfer of culture from one place to another place we use

a) Slant culture
b) Plate culture
c) Broth culture
d) Lyophilized culture

Q55. Spores are killed by

a) Autoclaving
b) Hot air oven
c) Gamma radiation
d) Flame sterilization

Q56. Separation of single colony is done by

a) Isolation
b) Inoculation
c) Incubation
d) All of these

Q57. Heavy metal salts used as disinfectants are-

a) Thiomersal nitrate b) Mercury

c) Lead d) All of these

Q58. Term Lyophilization indicates

a) Freeze drying b) Storage at 0c

c) Storage at -20 c d) None of these

Q59. Agar-Agar Is obtained from

a) Red Algae b) Brown algae

c) BGA d) Anabaena

Q60. Colony formation can be observed in liquid media broth.

a) True b) False

c) None of these

Q61. Which of these is indicator medium

a) Wilson and blair b) Mac conkey

c) Gelatin media d) Pikovskaya media

Q62. Blood agar medium is

a) Indicator medium b) Enriched medium

c) Differential media d) All the above

Q63. The culture medium whose exact composition is not known is called-

a) Simple b) Complex

c) Minimal d) Differential

Q64. Phenol coefficient indicates

a) Efficiencacy of a disinfectant

b) Conc of phenol

c) Amount of phenol

d) Efffecacy of phenol

Q65. Anaerobic microorganism are commonly grown in lab by

a) Pyrogallic acid over cotton

b) Oil over parafilm

c) Parafilm tap

d) Mineral oil over plate

Q66. Media used for *mycobacterium tuberculosis* in lab by

a) L J media b) Blood agar

c) Chocolate agar d) L B agar

Q67. Culture media commonly used for fungi cultivation is

a) Potato Dextrose Agar (PDA)

b) Nutrient Agar (NA)

c) Luria broth (LB)

d) Blood agar (BA)

Q68. Fermentation equipments are sterillised by

a) Moist heat b) Dry Heat.

c) Radiation d) All

Q69. Culture media used for *Leishmania* cultivation is

a) NNN media b) L J Media

c) PDA d) NA

Q70. Media used for *Staphylococci* cultivation is

a) Wilson and Blair b) LJ media

c) NNNN media d) NA

Q71. Kuffman white scheme is used to detect

a) *Staphylococci* b) *Streptococci*

c) *Mycobacterium* d) *Salmonella* spp

Q72. Gram positive bacteria which produce swarming on media is

a) *Proteus* b) *Bacillus*

c) *Corynebacterium* d) *Mycobacterium*

Q73. Which of the following is ionizing radaiation

a) U V b) Gamma

c) None of these d) Both a and b

Q74. Discountinous heating is called

a) Tyndalization b) Pasteurization

c) Boiling d) Lyophilization

Q75. Glassware are sterilized by

a) Dry heat b) Moist heat

c) Radiation d) Filteration

Q76. Father of antiseptic surgery

a) Leeuwenhook b) Joseph Lister

c) Koch d) None of these

Q77. Living cells are observed by——

a) Light microscope

b) Flourescent microscope

c) Electron microscope

d) Phase contrast Microscope

Q78. Dapi (diamino -2- phenylindole) is a fluorescent dye which stains cells

a) Blue b) Green

c) Red d) Yellow

Q79. Dic (Diffferential interface contrast) microscope gave 3d image of——

a) Eukaryotic Nucleus b) Prokaryotic cell

c) Mitochondria d) Plasmid

Q80. For biofilm observation we use——

a) Confoccal microscope b) Electron microscope

c) Phase contrast d) None of these

Q81. Transmission Electron Microscopy (TEM) is used for——

a) Internal cell structure b) External cell structure.

c) Flagella study d) Nucleus study

Q82. Resolving power of TEM is

a) 0.2 Nanometer b) 0.5 Nanometer

c) 0.2 Micrometer d) 0.1 Micrometer

Q83. Scanning Electron Microscopy (SEM) examines

a) 3D and external structurs b) Internal structures

c) Flagella d) Spore crystal bodies

Q84. Distance between the centre of the lens and the focal point is called

a) Focal length b) Resolving power

c) Focal point d) Numerical aperture

Q85. Immersion oil in microscope increases

a) Refractive index b) Focal lnght

c) Focal point d) Numerical aperture

Q86. The ability of microscope to distinguish between small objects that ate close together is called-

a) Resolution b) Refractive index

c) NA d) None

Q87. *Treponema palladium* is observed with

a) Dark field b) Bright field

c) Fluorescent microscope d) TEM

Q88. Bacterial compounds such as endospores, beta hydroxybutyrate are visualized by

a) Pahase contrast microscope

b) Light microscope

c) Lark field microscope

d) Fluorescent microscope

Q89. ……….. process by which internal and external structures of bacteria are fixed

a) Fixatation b) Smearing

c) Staining d) all

Q90. Methylene blue is a ————————-

a) Basic dye b) Acid dye

c) Neutral dye d) All of these

Q91. In gram stainig primary stain is—————-

a) Methylene blue b) Crystal violet

c) Saffranin d) Iodine

Q92. In gram stainig Iodine acts as—————

a) Modrant b) Decolourizer

c) Secondary stain d) Adjuvant

Q93. Acid fast stainig is used for——————-

a) Mycobacterium b) Treponema

c) Leishmania d) Staphylococcus

Q94. Capsules are stained by——————

a) Negative stainig b) Gram stainig

c) Acid fast staining d) Acid fast Stainig

Q95. Endospore stainig is done by—————-

a) Safranin b) Malachite green

c) Methylene blue d) Crystal violet

Q96. Flagella staining is done by

a) Ziel Nelson Method b) Gray method

c) Pararosaline method d) Both b & c

Q97. Resolution limit of light microscope is

a) 0.2 um b) 0.5 um

c) 0.1 um d) 0.7 um

Q98. Resolution of light microscope increases with the .... in the wavelength of the light it uses for illumination.

a) Increase b) Decreases

c) Both d) None

Q99. .... Processed TEM samples have the ability to see shape of organelles within microscope.

a) Freeze itching b) Glutarldehyde treated

c) None d) Both a & b

Q100. Temperature of liquid nitrogen is——————

a) -196 C b) -186

c) -203 d) -112

Q101. Process of burning municipal solid waste is called

a) Incineration b) Landfills

c) Composting d) Shredding

Q102. Which is a biodegradable waste

a) P bags b) Synthetic fiber

c) Food waste d) All

Q103. Waste dumped in soil is in

a) Composting b) Incineration

c) Landfills d) Shredding

Q104. Which of the below is not an idea behind solid waste management?

a) Control of waste generation

b) Storage and collection

c) Disposal

d) Stop waste generation

Q105. Processs of decomposition of biodegradable solid waste by earthworm is called-

a) Shredding b) Landfills

c) Vermicomposting d) Composting

Q106. Which is not a landfill method

A) Banglore method b) Area method

c) Depression method d) Trech method

Q107 ....is a liquid that passes through solid waste and extract suspended impurities from it

a) Lechate b) Sludge

c) Distilled water d) Municipal waste

Q108. Which of the following is not the municipal solid waste

a) Food waste b) Rubbish

c) Ashes d) All of these

Q109. .... Is the cutting and tearing of municipal solid waste-

a) Landfills b) Shredding

c) Pulverization d) Composting

Q110. In which method of composting, decomposition of anaerobic waste takes place

a) Banglore method b) Depression method

c) Trech method d) All of these

Q111. Biological method of waste disposal is

a) Composting b) Landfills

c) Shredding d) None of these

Q112. Composition of biogas is————

a) ch4-50-75%, co2-25-50% c n2-0-10%, h2-0-1% h2s-0-5%

b) ch4-60-75%, co2-45-50% c n2-0-12%, h2-0-2% h2s-0-1%

c) ch4-45-75%, co2-15-50% c n2-0-8%, h2-0-0.5% h2s-0-3%

d) ch4-35-75%, co2-25-30% c n2-0-5%, h2-0-2% h2s-0-3%

Q113. Use of living microorganisms to degrade environmental pollution is called –

a) Bioremediation b) Microremediation

c) Nanoremediation d) Both a & b

Q114. Which of the following called superbug

a) *Bacillus subtilis* b) *Pseudomonas putida*

c) *Bacillus denitrificans* d) None of these

Q115. Process of extracting metal from ore is called

a) Bioleaching b) Biofilteration

c) Bioextraction d) Microbial extraction

Q116. Which of the following microorganism involves in the removal of industrial wastes-

a) *Trichoderma* sp. b) *Aspergillus niger*

c) *Pseudomonas putida* d) All of above

Q117. Microorganisms remove metal by

a) Absorption and complextion

b) Adsorption and precipitation

c) Adsorption and volatilization

d) All

Q118. *Chlorella* sp are widely used for the removal of

a) Organic wastes b) Hydrocarbons

c) Heavy metals d) All

Q119. A non-directed Physico chemical interaction between heavy metal ions and microbial surface is called————

a) Biosorption b) Bioconversion

c) Biomining d) Biotransformation

Q120. ex-situ bioremediation involves————————

a) Removal of pollutants and collection at a place to facilitate microbial degradation

b) Degradation of pollutants by microbes directly

c) Degradation of microbes by genetically engineered microbes

d) Both a & b

Q121. Which of the following is not a method of composting—

a) Coimbatore method b) Banglore method

c) Indore method d) Hyderabad method

Q122. Biogas formation occurs under————

a) Aerobic conditions b) Anaerobic conditions

c) Both d) None of these

Q123. Municipal water supplies are purified by a process that contain four steps. The sequence of steps is-————

a) Sedimentation, settling or coagulation, rapid sand filteration and disinfection

b) Rapid sand filteration, settling or coagulation sedimentation and disinfection

c) Sedimentation, settling or coagulation, rapid sand filteration and disinfection

d) Rapid sand filteration and disinfection, sedimentation, settling or coagulation,

Q124. Cyclosporiasis is caused by————

a) Cyclospora b) Cryptosporidium

c) Giardia d) None of these

Q125. Canada goose is a major source of contamination of——

a) Giardia and Cryptosporidium

b) Cyclosporiasis

c) Both

d) None of these

Q126. When enteric indicator bacteria are not available in 100 ml of sample (water) it is called————

a) Potable b) Non potable

c) Autoclaved d) Sterilized

Q127. *E. coli*, entrobacter and *Klebsiella* are————

a) Facultative anaerobic b) Facultative aerobic

c) Obligate anaerobic d) Obligate aeorobe

Q128. TOC, COD& BOD provides different but complementary information on .... in a water sample———

a) Oxygen b) Hydrogen

c) Carbon d) Nitrogen

Q129. Primary treatment removes ....% pollution

a) 20-30% b) 10-15%

c) 15-25% d) 20-30%

Q130. Activated sludge and trickling filters are examples of

a) Aerobic secondary system

b) Anaerobic primary system

c) Anaerobic secondary

d) Tertiary

Q131. Phosphates and nitrogen are removed by

a) Precipitation and stripping

b) Boiling and volatilization

c) Precipitation only

d) Volatilization only

Q132. Constructed wetlands are used for the

a) Treatment of liquid wastes and bioremediation

b) Boiling and volatilization

c) Bioabsorption

d) Bioremediation

Q133. In septic tank what two processes occur————

a) Anaerobic liquefecation and digestion

b) Tert. Treatment

c) Both a & b

d) None of these

Q134. Giardiasis and cryptosporidiosis are———

a) Water borne disease b) Food borne disease

c) Air borne disease d) Contagious disease

Q135. Naegleria fourleri causes——

a) Amoebiasis b) Diarrhea

c) Fever d) Enteric fever

Q136. What ppm conc of chlorine in water ensures virus removal—

a) 0.6 ppm b) 0.8 ppm

c) 1 ppm d) 3 ppm

Q137. Legionellosis is caused by————

a) Bacteria b) Virus

c) Fungi d) Protozoa

Q138. Two basic techniques for microbiological assay are———

a) Agar diffusion & tube assay

b) Plate count

c) Haemocytometer

d) All of these

Q139. Vitamin B12 can be assayed by using———

a) *Ochromonas melhamensis*

b) *Lactobacillus lactis*

c) *Lactobacillus liechmanni*

d) All of these

Q140. *Lactobacillus fermenti* is used for the assay of

a) Vitamin B1
b) Vitamin B12
c) Vitamin C
d) Vitamin E

Q141. For streptomycin tube assay which organism is used———

a) *Bacillus subtilis*
b) *Staphylococcus aureus*
c) *E. coli*
d) *Bacillus cerus*

Q142. For tetracycline and penicillin G microbiogical assay which organism is used————

a) *Staphylococcus aureus*
b) *Staphylococcus epidermis*
c) *Bordetella bronchoiseptica*
d) None

Q143. For vancomycin assay which microorganism is used———

a) *B. cereus*
b) *B. subtilis*
c) *P. aureginosa*
d) *L. plantarum*

Q144. Keratinase enzyme is produce by————

a) Bacteria
b) Actinomycetes
c) Fungi
d) All

Q145. Amylase producing microorganism is————

a) *Pseudomonas*
b) *B. cereus*
c) *Geobacillus*
d) None

Q146. .... is the use of microorganism to access pollutio

a) Biological monitoring
b) Bioremediation
c) Bioleaching
d) Bioabsorption

Q147. Lichens are indicator of————

a) Air pollution
b) Water pollution
c) Soil pollution
d) None

Q148. Lichen is the association of————

a) Algae and fungi
b) Bacteria and virus
c) Fungus and plant
d) None of these

Q149. To combact petroleum pollution which bacteria is employed ———

a) *Acinetobacter baumannii* b) *P. putida*
c) *Leishmania* d) None of these

Q150. For chromium degradation which bacteria is used———

a) *P. mendocina* b) *P. putida*
c) *P. auriginosa* d) *Bacillus subtilis*

Q151. *Streptomyces rochei* degrade————

a) Chlorinated phenol b) Benzene
c) Hydrocarbon d) All

Q152. Colour reduction in distillery effluent is achieved with

a) *T. harzinum* b) *T. ressei*
c) *T. viride* d) *T. asperellum*

Q153. Which of these is not used for the complete mineralization of fungicides

a) *P. putida* b) *P. aureginosa*
c) *Serratia* d) *Bacillus*

Q154. Removal and killing of microbes is called————

a) Sterilization b) Disinfection
c) Pasteutization d) Lyophilization

Q155. To prevent the contamination of microscopes and surrounding areas, disenfect/clean used slides, prepared by student, with

a) 70% ethanol and lens paper
b) Acetone and lens paper
c) 5% methylene blue and lens paper
d) Water and lens paper

Q156. .... is the state which is completely free from microorganisms——

a) Asepsis b) Antibiotic
c) Sterilized d) Antisepsis

Q157. The method that ensure that surgical instruments are free from microorganisms are

a) Disinfected
b) Cleaned
c) Sterilised
d) Washed

Q158. Disinfectant that inhibit the growth of fungi is called———

a) Fungicidal
b) Fungisepsis
c) Fungistatic
d) None

Q159. The technique to avoid all microorganisms is————

a) Sterlization
b) Disinfection
c) Surgical sterilization
d) Disinfection Sterilization

Q160. What is thermal death time is

a) Time required to kill all cells at a given temperature
b) Temperature tha tkills all cells in a given time
c) Time and temperature needed to kill all cells
d) All of the above

Q161. Which method of sterilization has no effect on spores —

a) Drying
b) Hot air oven
c) Autoclave
d) None of these

Q162. The temperature and pressure required for autoclaving —

a) 121°C temp. and 1.5 lbs. pressure for 20 min.
b) 120°C temp. and 20 lbs. pressure for 30 min
c) 150°C temp. for 1 hr.
d) 130°C temp for 2 hr.

Q163. During preservation of microbes ——————

a) Their characteristics change
b) Their metabolism stop
c) Their metabolism continue
d) Their metabolism change

Q164. Microbes are preserved because ————

a) They produce valuable product

b) They are the only microbes producing product

c) They may require for longer period of time

d) All of the above

Q165. Cryopreservation is ————

a) Preservation at low temperature

b) Preservation at high temperature

c) Preservation with dehydration

d) Preservation in mineral oil

Q166. Which of the following method preserve microbes in active metabolic state?

a) Overlaying culture with mineral oil

b) Drying in vacuum

c) Lyophilization

d) Use of liquid nitrogen

Q167. Which of the following method of preservation can store the microorganisms for shorter period of time?

a) Growing culture with mineral oil

b) Drying in vacuum

c) Lyophilization

d) Storage in liquid nitrogen

Q168. In which of the following method of preservation are organisms dried over chemical?

a) Growing with mineral oil

b) Storage in sterile soil

c) Storage in normal saline

d) Drying in vacuum

Q169. Which of the following method is widely used for the preservation of microbes?

a) Drying in vacuum
b) Storage in sterile soil
c) Lyophilization
d) Storing organism in saline

Q170. Which of the following method used for the preservation of Spores?

a) Periodic transfer to fresh media
b) Storage in sterile soil
c) Storage in normal saline
d) Overlaying culture with mineral oil

Q171. Preserved culture is assessed for ....before use

a) Purity
b) Productivity
c) Viability
d) All of the above

Q172. For preservation cells should be in ______ phase of their life cycle

a) Log phase
b) Lag phase
c) Decline phase
d) Death phase

Q173. After a biohazard spill is covered with paper towels and disinfectant solution, it must sit for _________ minutes?

a) 5
b) 30
c) 60
d) 20

Q174. Which method of sterilization has no effect on spores

a) Drying
b) Hot air oven
c) Autoclave
d) None of these

Q175. Autoclaving is done of temperature and pressure which is

a) 121°C temp. and 1.5 lbs. pressure for 20 min.
b) 120°C temp. and 20 lbs. pressure for 30 min
c) 150°C temp.for 1 hr.
d) 130°C temp for 2 hr.

Q176. Commonly used filters are-————-

a) Filter paper b) Sieves

c) Nitrocellulose d) Filteration tubes

Q177. Pore size of nitrocellulose is——————-

a) 0.22 um b) 0.32 um

c) 0.21 um d) 0.19 um

Q178. Freeze drying or lyophilization is a method of————-

a) Maintenance and preservation

b) Culture transportation

c) Preservation

d) All of these

ANSWERS

1.	c	2.	b	3.	a	4.	b	5.	b
6.	d	7.	a	8.	a	9.	c	10.	c
11.	c	12.	b	13.	c	14.	b	15.	E
16.	E	17.	d	18.	c	19.	a	20.	d
21.	b	22.	d	23.	c	24.	c	25.	b
26.	b	27.	a	28.	c	29.	a	30.	b
31.	a	32.	b	33.	b	34.	b	35.	c
36.	a	37.	a	38.	d	39.	b	40.	c
41.	a	42.	c	43.	b	44.	b	45.	b
46.	a	47.	a	48.	a	49.	a	50.	a
51.	a	52.	a	53.	a	54.	a	55.	a
56.	a	57.	a	58.	a	59.	a	60.	b
61.	a	62.	b	63.	a	64.	a	65.	a
66.	a	67.	a	68.	a	69.	a	70.	a
71.	d	72.	a	73.	b	74.	a	75.	a
76.	b	77.	d	78.	a	79.	a	80.	a
81.	a	82.	a	83.	a	84.	a	85.	a
86.	a	87.	a	88.	a	89.	a	90.	a

91.	b	92.	a	93.	a	94.	a	95.	b
96.	b	97.	a	98.	b	99.	a	100.	a
101.	a	102.	c	103.	c	104.	d	105.	c
106.	a	107.	a	108.	c	109.	b	110.	a
111.	a	112.	a	113.	a	114.	b	115.	a
116.	c	117.	d	118.	c	119.	a	120.	a
121.	d	122.	b	123.	a	124.	a	125.	a
126.	a	127.	a	128.	c	129.	d	130.	a
131.	a	132.	a	133.	a	134.	a	135.	a
136.	a	137.	a	138.	a	139.	b	140.	a
141.	a	142.	a	143.	a	144.	d	145.	b
146.	a	147.	a	148.	a	149.	b	150.	a
151.	a	152.	c	153.	d	154.	a	155.	a
156.	a	157.	c	158.	c	159.	a	160.	a
161.	a	162.	a	163.	b	164.	d	165.	a
166.	a	167.	a	168.	d	169.	c	170.	b
171.	d	172.	a	173.	b	174.	a	175.	a
176.	c	177.	a	178.	a				

Unit-V

Microbial Biotechnology

Q1. Single cell protein (SCP) is the production of ?

a) Extracellular proteins

b) Fermentation of waste products

c) Intracellular proteins extraction

d) Metabolites

Q2. What do you mean by "Trophophase"?

a) Production of waste materials

b) Production of topical products

c) Production of primary metabolites

d) Production of secondary metabolites

Q3. What do you mean by "Idiophase"?

a) Production of waste materials

b) Production of topical products

c) Production of primary metabolites

d) Production of secondary metabolites

Q4. Which of the following does not have the property of production of secondary metabolites?

a) Filamentous fungi

b) Filamentous bacteria

c) Sporing bacteria

d) Enterobacteria

Q5. Is secondary metabolite useful?

a) True b) False

Q6. Microbial process is advantageous than chemical process.

a) True b) False

c) Both are false d) Both are true

Q7. Which of the following is an upstream process?

a) Product recovery b) Product purification

c) Media formulation d) Cell lysis

Q8. Which of the following is a downstream process?

a) Product recovery b) Screening

c) Media formulation d) Sterilization of media

Q9. Which of the following is not a product of fermentation?

a) Oxygen b) Carbon dioxide

c) Ethanol d) Lactate

Q10. Alcoholic fermentation is carried out by ——

a) *Lactobacillus* b) *Bacillus*

c) *Saccharomyces cerevisiae* d) *Escherichia coli*

Q11. Which of the following is not a probiotic?

a) Saccharomyces cerevisiae b) Escherichia coli

c) Lactobacillus d)none of these

Q12. Biofuels are products of fermentation————

a) True b) False

Q13. Biorefinery is not an example of fermentation————

a) True b) False

Q14. Fermentation which is carried by yeast is called———

a) Pyruvic fermentation b) Acrylic fermentation

c) Lactic acid fermentation d) Alcoholic fermentation

Q15. Considering fermentation at industrial level, microorganism *Bacillus* is used to form———

a) Ethanol b) Formic acid

c) Lactic acid d) Glycerol

Q16. Common example of fermented beverage product is——

a) Pickles b) Beer

c) Bread and buns d) Cheese and yogur

Q17. Applications of fermentation includes——————

a) Products b) Dairy products

c) Beverage products d) All of above

Q18. A concentrated solution of the limiting substrate is added in fed-batch at a rate?

a) Greater than the solution of the limiting substrate and the same medium used to establish the batch culture

b) Less than the solution of the limiting substrate and the same medium used to establish the batch culture

c) Equal to the solution of the limiting substrate and to the same medium used to establish the batch culture

d) Negligible to the solution of the limiting substrate and to the same medium used to establish the batch culture

Q19. Which of the following Fed-batch system is described as a fixed volume?

a) The same medium used to establish the batch culture

b) A solution of the limiting substrate

c) A concentrated solution of the limiting substrate

d) A very concentrated solution of the limiting substrate

Q20. What do you mean by "Quasi steady state"?

a) Cell concentration remains virtually constant

b) Cell concentration is virtually variable

c) Total biomass remains constant with time

d) Total biomass decreases with time

Q21. Quasi-steady state in fed batch is when?

a) Growth rate remains constant

b) Dilution remains constant

c) Growth rate changes variably

d) μ remains constant

Q22. What will be the condition when pH will be high?

a) Glucose is low

b) Glucose is equal with biomass concentration

c) Glucose is high with biomass concentration

d) Excess of glucose

Q23. Fed-batch culture is superior to conventional batch culture—

a) True b) False

Q24. Fed-batch culture is not used for substrate inhibition———

a) True b) False

Q25. Organisms in which phase are adapting to the new environment?

a) Lag phase b) Death phase

c) Exponential phase d) Stationary phase

Q26. Which type of media is used for fungi cultivation?

a) Non nutrient agar

b) Sabouraud's dextrose agar

c) MacConkey's agar

d) RPMI

Q27. Under continuous culture which of the following condition is applicable?

a) $\mu_{max} < D$ b) $\mu_{max} > D$

c) $\mu_{max} = D$ d) $\mu_{max}=0$

Q28. The growth rate of the cells will be greater than the dilution rate", Is this statement applicable for decrease in biomass production?

a) True b) False

Q29. An alternative of chemostat is turbidostat?

a) True b) False

Q30. What do you mean by wall growth in bioreactors?

a) Growth of cells in the wall

b) Consumption of paints by the cells from the coated walls

c) Biomass concentration is increased

d) Immobilized cells consumes substrate within reactors

Q31. The yield factor is proportional to the dilution rate?

a) True b) False

Q32. Mixed cultures can be maintained using chemostat cultures in ________

a) Fed-Batch culture b) Semi-Batch culture

c) Batch culture d) Continuous culture

Q33. Which of the following is not an advantage of continuous culture?

a) Can be used for different reactions every day

b) Little risk of infection or strain mutation

c) Long growth periods of subtrates/microbes

d) Eliminating the inherent down time for cleaning and sterilization

Q34. Which of the following is not a disadvantage of continuous culture?

a) Long growth periods of subtrates/microbes

b) Maintenance of mixed cultures

c) Requires feed-batch culturing

d) Viscosity of mixture for filamentous organisms

Q35. Continuous addition of sugars in “Fed-batch” fermentation is done to —

a) Produce methane b) Purify enzymes

c) Degrade sewage d) Obtain antibiotics

Q36. Name the phase which is a period of adaptation of the cells to the new environment.

a) Lag phase b) Log phase

c) Exponential phase d) Stationary phase

Q37. In chemostat, constant cell concentration is maintained?

a) True b) False

Q38. Which of the following is used to grow bacterial cultures continuously?

a) Haemostat

b) Chemostat

c) Bacteria cannot be grown in continuous culture

d) Thermostat

Q39. Pasteur effect discovered in 1857, is

a) Inhibiting effect of oxygen on the fermentation process

b) Aerating yeasted broth causes yeast cell growth to decrease, while conversely, fermentation rate increases

c) A and B

d) All of these

Q40. The final electron acceptor in lactic acid fermentation is

a) Oxygen b) Lactic acid

c) Pyruvate d) NAD

Q41. Identify the correct sequence during the industrial production of substances

a) Inoculation, screening, fermentation, downstream processing, removal of waste

b) Screening, Inoculation, fermentation, downstream processing, removal of waste

c) Fermentation, screening, inoculation, removal of waste, downstream processing

d) Fermentation, inoculation, inoculation, removal of waste, downstream processing

Q42. Production of ____________ occurs in the fermentation of grains

a) Must b) Brine

c) Lactic acid d) Carbon dioxide

Q43. ____________ is the detrimental effect of microorganisms in the food

a) Generation of flavor b) Process of fermentation

c) Spoilage of food d) All of the above

Q44. Which is not an advantage of the fermented food?

a) Makes the food more digestible

b) Increase storage life

c) Synthesize vitamins

d) Decrease intestinal microflora

Q45. Which is not fruit or vegetable based fermented product

a) Wine b) Beer

c) Vinegar d) Sauerkraut

Q46. Large vessel containing all the parts and condition necessary for the growth of desired microorganisms is called

a) Bio reactor b) Auto reactor

c) Impeller d) None of these

Q47. For thorough mixing of medium and inoculum the part of fermenter useful is

a) Shaft b) Headspace

c) Impeller d) Sparger

Q48. In fermenter the top portion left without broth is called

a) Shaft b) Headspace

c) Impeller d) Sparger

Q49. The factors affecting KLa values in fermentation vessels are the following except

a) The air flow rate employed

b) The degree of agitation

c) The rheology properties of the medium

d) The presence of chelating agents

Q50. Antifoam agents collapse the foam and thus increase the .... transfer rate of the fermentation medium

a) Carbon dioxide b) Nitrogen

c) Oxygen d) Hydrogen

Q51. The degree of agitation affects in fermentation process except ___________

a) It increase the contact time for bubbles in the medium

b) It plays a vital role in the oxygen transfer rate in agitated fermenter

c) It influence coalesces of air bubbles

d) It decreases thickness of liquid film at gas-liquid interface

Q52. Medium Rheology is ________________

a) Medium flow characteristics

b) Air flow characteristics

c) Mass flow characteristics

d) Liquid flow characteristics

Q53. Bubble column reactor cannot be used for _______________

a) Highly viscous medium b) Low viscous medium

c) Solid state medium d) Liquid state medium

Q54. The dynamic method of gassing out procedure ——-

a) Increase the supply of air to the fermentation

b) Stopping the supply of air to the fermentation

c) Changing the supply of air to the fermentation

d) None of these

Q55. Protein content called novel is produced by using———

a) Bacteria

b) Virus

c) Fungi

d) Micro-organisms

Q56. Protein content which is extracted from mixed or pure cultures of yeasts, bacteria, algae and fungi is called

a) Triple cell protein

b) Single cell protein

c) Double cell protein

d) Tetra cell protein

Q57. Technique of SCP is introduced by———-

a) Gregor Mendel

b) Louis Pasteur

c) Professor Scrimshaw

d) Ian Wilmot

Q58. By using single-cell protein, amount of protein that can be produced by algae grown in ponds (per acre) is

a) 20 tons

b) 30 tons

c) 40 tons

d) 50 tons

Q59. Substrate used by microorganisms to produce single-cell proteins includes

a) Methane gas

b) Industrial wastes

c) Agricultural wastes

d) All of above

Q60. Bacteriocins are———

a) Bacteriocins are proteinaceous or peptidic toxins produced by bacteria to inhibit the growth of similar or closely related bacterial strain(s)

b) Chemical compounds

c) Bacterially produced heat stable antimicrobial pepetide

d) None of these

Q61. Which of these statements is false?

a) Lactobacillus increases the incidence of Diarrrhea

b) Lactobacillus decreases the incidence of Diarrrhea

c) Both a & b

d) None of these

Q62. Antibiotic-associated diarrhea can be reduced the most with:

a) Antibiotic alone

b) Probiotic consumption during antibiotic administration

c) Antibiotic with hormones

d) Antibiotic alone

Q63. Probiotics can kill pathogens by——-

a) Disrupting their cell wall

b) Denaturing enzymes

c) Inhibiting protein synthesis

d) None of these

Q64. Prebiotic fermentation results in a selective stimulation of—

a) Lactic acid bacteria b) Lactose induction

c) Casein induction d) All of these

Q65. Benefits of prebiotic fermentation include:

a) An anti-inflammatory effect due to butyrate production

b) Butyrate production

c) Lactic acid bacteria

d) None of these

Q66. Probiotics can kill pathogens by

a) Disrupting the pathogen,s cell wall

b) Inhibiting cell wall synthesis

c) Inhibiting protein synthesis

d) All of the above

Q67. Probiotic are————

a) Cancer inducing microbes

b) Safe antibiotics

c) New kind of food allergens

d) Live microbial food supplement

Q68. There are more gut bacteria than the total number of cells in the body

a) True
b) False

c) Both a & b
d) None of these

Q68. Which of the following is a factor that affects the storage stability of food?

a) Type of raw material used

b) Quality of raw material used

c) Method/effectiveness of packaging

d) All of the mentioned

Q69. Which of the following sentence is true with respect to food storage/preservation?

a) Each food type has a potential storage life

b) The mechanical abuse that food has received during storage/distribution affects its storage stability

c) Both of the mentioned

d) None of the mentioned

Q70. Statement 1: Foods of plant origin can be used as additives for food preservation.

Statement 2: Dry fruits and seeds are the most important higher plant structures used as food.

a) True, False
b) True, True

c) False, False
d) False, True

Q71. Which of the following statement with respect to food preservation is true?

a) Leafy vegetables perish fast due to their high moisture content

b) Cereals have the highest requirements of moisture and soil types

c) Cereal can be grown with less labour and yield of food is high

d) All of the mentioned

Q72. Cereals are the major source of food in the world

a) True b) False

Q73. Cereals are a major source of carbohydrates

a) True b) False

Q74. Statement 1: Majority of the fish have more proteins than water.

Statement 2: Whole milk has more water than fat.

a) True, False b) True, True

c) False, False d) False, True

Q75. The storage of prepared food in ____ areas in the ____ of oxygen creates conditions for ____ Which option best fits the above sentence?

a) Cold, presence, Purification

b) Warm, absence, Putrefaction

c) Cold, presence, Putrefaction

d) Warm, absence, Purification

Q76. Which of the following is NOT a process wherein the food becomes toxin before ingestion?

a) Botulism b) Staphylococcus

c) Bacterial Intoxication d) Bacterial infection

Q77. Statement 1: Botulism is more dangerous than Staphylococcus

Statement 2: Botulism is encountered by humans only if they've eaten the toxin. The organism in itself is no harm. Staphylococcus needs air and grows on warm food only

a) True, False b) True, True

c) False, False d) False, True

Q78. Statement 1 : Foreign objects entering food is called physical contamination of food.

Statement 2 : Controllingmoisture is the only precaution to be taken to prevent food contamination.

a) True, False b) True, True

c) False, False d) False, True

Q79. Statement 1: Processing contaminants, are the contaminants that are generated during the processing of food and hence are hard to control.

Statement 2 : Packaging materials also cause contamination.

a) True, False b) True, True

c) False, False d) False, True

Q80. Diarrhea, vomiting and sever abdominal cramps shows their sign in

a) Food poisoning b) Constipation

c) Heart diseases d) Muscle cramps

Q81. Bacteria which is present in raw or undercooked meat, eggs, sea food and unpasteurized milk is

a) *E. coli* b) *Salmonella*

c) *Staphylococcus* d) *Cyano bacteria*

Q82. During baking process of bread ethanol is

a) Absorbed b) Evaporated

c) Condensed d) Reacted

Q83. The staphylococcal intoxication refers to presence of——-

a) An enterotoxin b) Neurotoxin

c) Mycotoxin d) All of these

Q84. A bacterial food intoxication refers to

a) Ilness caused by oresence of pathogens

b) Food borne illness caused by the presence of a bacterial toxin formed in food

c) Both a & b

d) None of these

Q85. The method of successful treatment of botulism prior to appearance of botulism symptoms involve administration of

a) Antibiotic b) Analgesic

c) Antitoxin d) Antipyretic

Q86. Botulism is caused by the presence of toxin developed by——————

a) *Clostridium tyrobutyricum*

b *Clostridium sporogenes*

c) *Clostridium botulinum*

d) None of these

Q87. Which of the following is true to prevent botulism from smoked fish?

a) The fish should be heated at its coolest part to at least 82°C for 30 min during or after smoking

b) The fish should be heated at its coolest part to at least 82°C for 30 min during or after smoking

c) Good sanitation should be maintained throughout production and handling

d) All of the above

Q88. *Salmonellois* is caused by the——————

a) Enterotoxin of *Salmonella* spp

b) Endotoxin of *Salmonella* spp

c) Neurotoxin of *Salmonella* spp

d) Exoenterotoxin of *Salmonella* sp

Q89. Group I *C.botulinum* strains generally includes in——

a) All types of strains (proteolytic) A, B and F

b) All types of strains (non-proteolytic) E and F

c) All types of strains (proteolytic) C, D and F

d) One of the above

Q90. Arrange the potencies of toxin in descending order of the following canned foods

a) Peas>string beans>spinach> corn

b) Corn> string bean>peas>spinach

c) Corn> spinach> peas>string beans

d) Corn> peas>string beans>spinach

Q91. The milk streptococci produce acetoin that gets spontaneously oxidized yielding a flavorings agent (responsible for aroma of butter) that is

a) Acetone b) Acetyl coA

c) Butyric acid d) Diacetyl

Q92. The botulism intoxication occurs due to

a) An enterotoxin b) Neurotoxin

c) Mycotoxin d) All of these

Q93. The *Bacillus cereus* causes gasteroenteritis by the production of an exoenterotoxin which is released in food as a result of

a) Cell growth b) Cell autolysis

c) Cell permeation d) Cell permeation

Q94. Staphylococcal intoxication is caused by the toxin in the food from

a) *Staphylococcus aureus* b) *S. cerevisiae*

c) *S.thermophillus* d) None of these

Q95. Regarding methanol poisoning assertion: Administration of ethanol is one of the treatment modalities

Reason: Ethanol inhibits alcohol dehydrogenase

Please select the most correct option from the following:

a) Both assertion and reason are true, and the reason is the correct explanation for the assertion

b) Both assertion and reason are true, and the reason is not the correct explanation for the assertion

c) Assertion is true, but the reason is false

d) Assertion is false, but the reason is true

Q96. In methyl alcohol poisoning, there is central nervous system depression, cardiac depression and optic nerve atrophy. These effects are produced due to

a) Formaldehyde and formic acid

b) Acetaldehyde

c) Pyridine

d) Acetic acid

Q97. Phosphine liberated in the stomach in aluminum phosphide poisoning is toxic to all except

a) Lungs b) Kidneys

c) Liver d) Heart

Q98. Paraquat poisoning causes————

a) Renal failure b) Cardiac failure

c) Respiratory failure d) Multiple organ failure

Q99. Which of the following microbe is used in the production of blue cheese?

a) *Streptococcus thermophilus*

b) *Lactobacillus bulgaricus*

c) *Penicillium roqueforti*

d) *Rhizopus stolonife*

Q100. Pickled cucumber is made from fermented salt-stock pickles.

a) True   b) False

Q101. Bacterial cell grown on hydrocarbon wastes from the petroleum industry are a source of ____________

a) Carbohydrates   b) Proteins

c) Vitamins   d) Fats

Q102. Yeast-cell crops harvested from the vats is used to produce which of the following compounds?

a) Alcoholic beverages   b) Enzymes

c) Antibiotics   d) Organic acids

Q103. How many tons of protein can be produced by algae grown in pond in a year?

a) 1000   b) 1.0

c) 50   d) 20

Q104. What is the range of protein content in yeast cells?

a) 69%   b) 12-15%

c) 20-40%   d) 40-50%

Q105. Which of the following microorganism have a high vitamin content?

a) Bacteria   b) Yeast

c) Algae   d) Protozoa

Q106. The principal microorganism for yogurt is ___________

a) *Streptococcus thermophilus*

b) *Leuconostoc citrovorum*

c) *Lactobacillus acidophilus*

d) *Streptococcus lactis*

Q107. Which of the following products have higher acidity and lacks aroma?

a) Cultured buttermilk b) Cultured sour cream

c) Bulgarian milk d) Acidophilus milk

Q108. Shredded cabbage is the starting product for which of the following fermented food?

a) Sauerkraut b) Pickles

c) Green olives d) Sausage

Q109. Most spoilage bacteria grow at————

a) Acidic pH b) Alkaline pH

c) Neutral pH d) any of the pH

Q110. The microbiological examination of coliform bacteria in foods preferably use———-

a) MacConkey broth b) Violet Red Bile agar

c) Eosine Methylene blue agar d) All of these

Q111. Which of the following acid will have higher bacteriostatic effect at a given pH?

a) Acetic acid b) Tartaric acid

c) Citric acid d) Malic acid

Q112. What are the intrinsic factors for the microbial growth?

a) pH

b) Moisture

c) Oxidation-Reduction Potential

d) All of these

Q113. NaCl can act as————

a) Antagonist at optimal concentrations

b) Synergistically if added in excess of optimum level

c) Both (a) and (b)

d) None of the above

Q114. Which of the following is responsible for a musty or earthy flavor?

a) Actinomycetes b) Flavobacterium

c) Both (a) and (b) d) *Pseudomonas syncyanea*

Q115. Molds causing spoilage of eggs include species of

a) *Cladosporium* b) *Mucor*

c) *Thamnidium* d) All of these

Q116. Vacuum packagedmeats are spoiled by

a) *Bthermosphacta* b) *Lactobacilli*

c) Both (a) and (b) d) None of these

Q117. More spoilage of eggs is caused by

a) Bacteria than molds b) Molds than bacteria

c) Synergistically d) None of these

Q118. Unopened, sliced bacon, packaged in material somewhat highly resistant to oxygen permeability, is spoiled mostly by

a) *Lactobacilli* b) *Micrococci*

c) *Fecal Streptococci* d) *Molds*

Q119. Pink rots in eggs is caused by the strains of————

a) *Pseudomonas*

b) *Pseudomonas fluorescens*

c) Species of *Micrococcus or Bacillus*

d) One of the above

Q120. Beef hams are made spongy by species of———

a) *Rhodotorula* b) *Bacillus*

c) *Pseudomonas* d) Red *Bacillus*

Q121. The red color of meat, called its bloom, may be changed to shades of green, brown or grey as a result of production of Oxidizing compounds by bacteria. Which of the following species are reported to cause the greening of sausage?

a) *Lactobacillus* b) *Leuconostoc*

c) Both (a) and (b) d) *Pseudomonas*

Q122. Yellow discolorations in meat are caused by bacteria with yellow pigments, usually species of

a) *Micrococcus* b) *Flavobacterium*

c) Both (a) and (b) d) *Pseudomonas syncyanea*

Q123. Which of the following (s) is/are responsible for the green patches on the surface of meats under aerobic conditions?

a) *P. expansum* b) *P. asperulum*

c) *P.oxalicum* d) All of these

Q124. Which of the following microorganisms grow in beef at a temperature of 15°C and above?

a) *Micrococci* b) *Pseudomonas*

c) Both (a) and (b) d) *Lactobacillus*

Q125. Green rots in eggs is chiefly caused by———

a) *Pseudomonas fluorescens*

b) *Micrococcus* or *Bacillus species*

c) *Molds* or *yeasts*

d) All of the above

Q126. Black rots in eggs is most commonly caused by——-

a) Species of *Proteus*

b) Species of *Micrococcus* or *Bacillus*

c) Molds or yeasts

d) All of these

Q127. Which of the following microorganism spoil poultry in polyethylene bags?

a) *Pseudomonas-achromobacter*

b) *Alcaligenes*

c) Hetero-fermentative species

d) Catalase negative bacteria

Q128. Black spot in meat are produced mainly by

a) *Cladosporium*

b) *Thamnidium*

c) *Mucor*

d) *Rhizopus*

Q129. Which of the cause(s) has/have been suggested for the chill rings in sausage?

a) Oxidation

b) Production of organic acids or reducing substances by bacteria

c) Excessive water

d) All of the above

Q130. Lactic acid bacteria in meats may be responsible for———-

a) Slime formation at the surface or within especially in presence of sucrose

b) Production of green discoloration

c) Souring

d) All of the above

Q131. Colorless rots in eggs is caused by————-

a) *Pseudomonas alcaligenes*

b) *Pseudomonas fluorescens*

c) both (a) and (b)

d) none of these

Q132. Iced, cut up poultry often develops a slime i.e. accompanied by an odor describe as tainted, acid, sour or dishraggy. This defect is chiefly by the species of————

a) *Pseudomonas*

b) *Alcaligenes*

c) Hetero-fermentative species

d) Catalase negative bacteria

Q133. Which of the following is recommended for the improvement in flavor during ageing of beef under controlled conditions?

a) *Thamnidium* b) *Rhizopus*

c) *T.elegans* d) *M.mucedo*

Q134. White spot in meat is formed due to presence of———-

a) *P.expansum* b) *Sporotrichum carnis*

c) Both (a) and (b) d) *P.oxalicum*

Q135. The chief spoilage organisms on smoked fish are———

a) Molds b) Bacteria

c) Both (a) and (b) d) Fungi

Q136. A musty or muddy odor of the fish is attributed to———-

a) The growth of Streptomyces species in the mud at the bottom of the body of water

b) The mud at the bottom of the body of water

c) The growth of Pseudomonas species in the mud at the bottom of the body of water

d) None of the above

Q137. In chilled shrimp __________ is chiefly responsible for spoilage.

a) *Achromobacter*

b) *Pseudomonas*

c) *Micrococcus* or *Bacillus* species

d) *Molds* or *Yeasts*

Q138. Marinated (sour pickled) fish should not have spoilage problems unless

a) The acid content is very high

b) The acid content is low enough

c) The acid content is moderate

d) None of the above

Q139. The predominant kind of bacteria causing spoilage in fish at chilling temperature is————

a) *Pseudomonas* spp b) *Micrococcus*

c) *Bacillus* d) None of these

Q140. The bacteria most often involved in the spoilage of fish are— ————

a) Part of the natural flora of the external slime of fishes and their intestinal contents

b) Part of the natural flora of the internal slime of fishes only

c) Both (a) and (b)

d) None of the above

Q141. The red or pink color of the fish is generally caused from the growth of

a) *Sarcina*

b) *Micrococcus* or *Bacillus* species

c) Molds or yeasts

d) All of these

Q142. Yeast are most likely to grow in frozen fruits during——

a) Slow thawing b) Refrigeration

c) Ambient temperature d) None of these

Q143. Watery soft rot is found mostly in————-

a) Fruits b) Vegetables

c) Cereals d) All of these

Q144. Black mold rot is caused by————

a) *A. flavus*
b) *A. niger*
c) *Trichoderma*
d) *Trichothecium roseum*

Q145. Concentrate of fruits and vegetable juices———

a) Favor the growth of *a. Niger* and *a. Flavus* species
b) Favour the growth of yeast and of acid and sugar tolerant *Leuconostoc* and *Lactobacillus* species
c) Favor the growth of saprophytic bacteria
d) None of the above

Q146. The predominant micro-organisms in frozen foods are——

a) Bacteria
b) Micro-coccus
c) Yeast and moulds
d) None of these

Q147. Bacterial soft rot is caused due to————-

a) Fermentation of pectin
b) Fermentation of sugars
c) Formation of ketones
d) Formation of amino acids

Q148. The micro-organism which is present in both fresh and frozen juices is

a) *E. coli*
b) *Entereobacter aerogenes*
c) *P. chrysogenum*
d) None of these

Q149. Flat sour spoilage of acid foods is caused by————-

a) *B. coagulans*
b) *B. pepo*
c) Both (a) and (b)
d) *B. stearothermophillus*

Q150. The ability of *B. coagulans* to grow in tomato juice depends upon the

a) Number of spores present
b) The availability of oxygen
c) The pH of the juice
d) All of these

Q151. Which of the following strain(s), that grow in the jellies and candid fruits, are able to grow in sugar concentration upto 67.5%?

a) *Aspergillus*
b) *Penicillium*
c) *Citromyces*
d) All of these

Q152. Medium acid foods with a pH between 5.3 and 4.5 are likely to undergo

a) Flat sour spoilage b) Putrefaction

c) Both (a) and (b) d) Spoilage

Q153. Canned sweetened condensed milk may become thickened by————

a) *Bacillus* species b) *Clostridium* species

c) *Micrococcus* species d) *Saccharomyces species*

Q154. The canned acid foods such as pineapple, tomatoes and pears have been found spoiled by *C. pasteurianum*. Such spoilage is more likely when the pH of the food is

a) >4.5 b) <4.5

c) >5.3 d) <5.3

Q155. Which of the following species can cause bitterness, acidity and curdling in canned milk, cream and evaporated milk?

a) *Bacillus* b) *Clostridium*

c) Both (a) and (b) d) *Saccharomyces*

Q156. Canned poultry is more often spoiled by————-

a) Putrefactive than by *Saccharolytic Clostridia*

b) *Saccharolytic Clostridia* than by *putrefactive*

c) *Saccharomyces* species

d) *Micrococcus* species

Q157. Canned meat and fish exhibit spoilage by————

a) *Bacillus* species b) *Clostridium* species

c) Both (a) and (b) d) None of these

Q158. In most fruit juices the major portion of total soluble solids is consists of

a) Salt b) Sugar

c) Vitamin c d) Mineral

Q159. Most commonly processed juice is————

a) Apple juice b) Orange juice

c) Pineapple juice d) Strawberry juice

Q160. Which bacteriumis capable of withstanding doses of radiation. That are several times higher than human cells can tolerate?

a) *Staphylococcus aureus*

b) *Deinococcus radiodurans*

c) *Escherichia coli*

d) *Saccharomyces cerevisiae*

Q161. Which is not an indigenous microbe used for bioremediation?

a) *Piscirikettsis salmonis* b) *E. coli*

c) *Phanerochaete sordida* d) *Pseudomonas aeruginosa*

Q162. Ananda Chakrabarty received the first U.S. patent for a GM organism. This organism was:

a) A transgenic mouse expressing the growth hormone gene

b) Cloned *E.coli*

c) The Glofish

d) *Pseudomonas* engineered to degrade petroleum

Q163. During which stage of wastewater treatment methanogenic microbes are most important?

a) Primary treatment b) Sludge digestion

c) Biological oxidation d) Secondary treatment

Q164. Anaerobic bacteria often play important roles in bioremediation. Which of the following is not an electron acceptor used by anaerobes during biodegradation reactions?

a) CO_2 b) NO_3

c) Fe (III) d) H_2O

Q165. Bioaugmentation is a process that involves:

a) Using plants for bioremediation

b) Bioventing

c) Sludge removal

d) Adding microbes to a cleanup site

Q166. Which bioremediation approach involves mixing of contaminated soil with water,carbon dioxide,and fertilizers in a bioreactor to stimulate biodegradation?

a) In situ hybridization

b) Slurry-phase bioremediation

c) Biopile treatment

d) In situ bioremediation

Q167. Which bioremediation approach involves using plants to degrade pollutants?

a) Biopile
b) Phytoremediation
c) Composting
d) Land farming

Q168. During which stage of wastewater treatment is the primary effluent aerated to allow for biodegradation by aerobic microbes?

a) Sedimentation
b) Secondary treatment
c) Sludge digestion
d) Disinfection

Q169. Which cleanup approach involves removing groundwater or soil from its natural setting to allow bioremediation?

a) *In situ* bioremediation
b) *Ex situ* bioremediation
c) Bioaugmentation
d) Phytoremediation

Q170. Xenobiotics" are ______________.

a) Synthetic organic compound found in nature

b) Organic compound

c) Inorganic compounds

d) Natural compounds

Q171. After being applied to soil, organic compounds may be destroyed by which of the following processes?

a) Biological degradation b) Chemical degradation

c) Both a & b d) None of these

Q172. Phytoremediation can clean up polluted soils by using——

————

a) Plants to take up and accumulate the pollutant so that it can be removed when the plant is harvested

b) Removal of pollutants by fungi.

c) Removal of pollutants by bacteria

d) All of these

Q173. Bioremediation can clean up polluted soils by————-

a) Adding nutrients to stimulate the activity of certain soil bacteria

b) Inoculating the soil with certain bacteria that can degrade toxic organic compounds

c) Using plants to stimulate soil microbial activity by exuding energy compounds into the rhizosphere

d) None of these

Q174. Citric acid is produced by————

a) *Aspergillus niger* b) *Pseudomonas putida*

c) *B.cerus* d) *Trichoderma viride*

Q175. Kojic acid is produced by————

a) *Aspergillus flavus* b) *Aspergillus niger*

c) *B. subtilis* d) *P. putida*

Q176. Ethanol production from lactose is occurred by————

a) *Kluyveromyces fragilis* b) *Sachhormyces cerevesie*

c) *Acetobacter acetie* d) All of these

Q177. Lysine production is carried out by————-

a) *Corynebacterium glutamicum*

b) *Corynebacterium dipthereai*

c) Both a & b

d) Nonc of these

Q178 .... Is added to maximize the Penicillin G production

a) Phenylacetic acid b) Acetic acid

c) Benzoic acid d) All of these

Q179. Streptomycin is produced by—————-

a) *Streptomyces griseus* b) *Strptomyces* spp

c) *Penicillium notatum* d) None of these

Q180. Dextrans are used as————-

a) Blood Expanders

b) Blood absorbents

c) Blood expenders & absorbents

Q181. Secondary metabolite production occurs during————-

a) Stationary growth phase b) Log phase

c) Lag Phase d) Death phase

Q182. Metabolite formed during growth phase are—————

a) Primary Metabolites b) Secondary metabolites

c) Bboth a & b d) None of these

Q183. Procees of transferring small scale equipment to large scale equipment is called

a) Scale up b) Scale down

c) Both a&b d) None of these

Q184. For vit B12 production microbial strain used are————

a) *Propionibacterium* and *Pseudomonas*

b) *P. seudomonas*

c) *B. subbtilis*

d) All of these

Q185. Fungus *Ashbrya gossypium* produces large amounts of —

a) Riboflavin b) Vit. C

c) Vit. D d) Vit A

Q186. Aspartame is————

a) Non nutritive Sweetener b) Sweeetner

c) Nutritive Sweetner d) None of these

Q187. Which Amino acis is used as flavor enhancer————

a) Glutamic Acid b) Aspartic Acid

c) Lysine d) All of these

Q188. *Bravicaterium Flavum* is used for the production of—

a) Lysine b) Glysine

c) Serine d) None of these

Q189. A wine in which Brandy or some other alcoholic spirit is added after fermentation is called

a) Fortified wine b) Un fortified wine

c) Both d) None of these

Q190. Sherry and port are the examples of————

a) Dry wines b) Sweeet wines

c) Fotified wines d) None of these

Q191. Procee of making alcoholic beverages from malted grains is called

a) Brewing b) Encology

c) Fermentation d) None of these

Q192. Key genera of acetic acid production bacteria are———

a) Acetobacter b) Gluconobacter

c) Both a&b d) None of these

Q193. In mushroom farms which species of mushrooms is commonly grown

a) *Agaricus bisporus* b) *Chlorella*

c) *Gyromitra esculenta* d) *Cantharellus cibarius*

Q194. In yogurt which two type of bacteria are present———

a) *S. thermophilus* & *L. bulgaricus*

b) *S. thermophillus* & *A. aceteti*

c) Both a & b

d) None of these

Q195. *Bifidobacterium* is generally used in————-

a) Milk fermentation b) Alcohol fermentation

c) Cheeese production d) All of these

Q196. Kefir is produced by—————

a) Yeast- lactic acid bacteria b) Yeast alone

c) Both a & b d) None of these

Q197. Villli is produced by———-

a) Yeast- lactic acid bacteria

b) Yeast-Geotrichium

c) *Geotrichium candidum* and lactic acid bacteria

d) None of these

Q198. Soft cheese are ripened for—————

a) 1-5 months b) 2-8 months

c) 1-2 months d) None of these

Q199. Period of ripening for hard cheese is-

a) 1-12 months b) 3-12 months

c) 4-10 months d) 8-10 months

Q200. Process of wine making is called—————

a) Enology b) Encology

c) Alchohol fermentation d) None of these

Q201. In pickle formation which bacteria plays an important role

a) *L. plantarum* b) *L. brevis*

c) Both a & b d) Non of these of these

Q202. Primary microorganisms contribute to Sauerkraut or sour cabbage are-

a) Leuconostoc *mesenteroides*

b) Lactobacillus *plantarum*

c) Both a & b

d) None of these

Q203. Microoorganisms such as *Bifidobacterium* & *Lactobacillus* are primarily used in

a) Prebiotics b) Probiotics

c) SCP d) None of these

Q204. A countinous bioreactor in which only the flow rate is used to control the rate of cell or product formation is called

a) Chemostat b) Turbidostat

c) Countinous culture d) All of these

Q205. Lowest yield of ATP is in

a) Fermentation b) Respiration

c) Photosynthesis d) None of these

Q206. When two populations compete for a single growth limiting substrate in a continuous fermenter which organism would not be washed out

a) Organism maintaining the lowest substrate concentration

b) Organism maintaining the high substrate concentration

c) Both a & b

d) None of these

Q207 .... is an efficient method of producing energy from biomass

a) Fermentation b) Respiration

c) Photosynthesis d) All of the above

Q208. *S. cerevisise* is type of————

a) Yeast b) Candida

c) SCP d) None of these

Q209. Which of the following crop is not used for the production of ethanol

a) Oil crop b) Corn starch

c) Cane sugar d) None of these

Q210. Which enzyme is used for juice clarification————-

a) Amylase b) Pectinase

c) Starch d) All of these

Q211. Which of the following was the first amino acid to be produced coomercially

a) L-Glysine b) L-lysine

c) D-Lysine d) D-Glysine

Q212. A countinous bioreactor in which only the rate is used to control the rate of cell or product productivity is called

a) Chemostat b) Turbidostat

c) Both a & b d) None of these

Q213. Cuntinous culture are not widely used in industry because——-

a) Not suitable for secondary metabolite production and chances of contamination or mutation are high

b) Cross contamination chances are high

c) Handling is very tuff

d) All of the above

Q214. Which material is used as bioplastic————-

a) Polyhydroxybutyrate b) Hydroxybutyrate

c) Both a&b d) None of these

Q215. Trickling filters are used for———-

a) Waste water treatment b) Antibiotic production

c) Amino acid production d) All of these

Q216. Importance of yeast extract in industrial fermenter is——

a) Act as vitamin or micronutrient source

b) Act as vitamin source

c) Act as growth enhancer

d) All of these

Q217. Soy meal, peptone and tryptone are used as source of——

a) Nitrogen source b) Amino acid source

c) Carbohydrate source d) None of these

Q218. Air lift fermentor uses—————-

a) Differential density for mixing purpose

b) Density for mixing

c) Uplifting movement for mixing

d) All of these

Q219. Spray dryer work on the principle of—————-

a) Adiabatic drying b) Diabatic drying

c) Simple drying d) None of these

Q220. Protected fermentation uses————-

a) Pasteurized media with low pH

b) Pasteurized media with high pH

c) Unpasteurized media

d) All of these

Q221. Molases and corn steep liquor are usually used as………. Source for the large scale industrial fermentation

a) Carbon b) Nitrogen

c) Hydrogen d) None of these

Q222. Importance of tryptophan in microbiology is as————

a) To provide C

b) To provide N

c) To produce lsogeny for microbes

d) All of these

Q223. Marmite an important coomercial product is a————-

a) Fermentated by product b) Fermentated product

c) Cheese d) None of these

Q224. Process of production of EtOH by yeast cells under high conc of glucose rather than producing biomass by TCA is described as

a) Crabtree effect b) Glyoxlate cycle

c) Pyruvate cycle d) None of these

Q225. A bioreactor in which fresh medium is countinously added and waste material is removed at the same rate is called

a) Chemostat b) Turbidostat

c) Countinous culture d) None of these

Q226. Pyrethrin is obtained from————

a) *Chrysantheum cinerarifolium*

b) *Chrysantheum chrysogenum*

c) Rosa sinenesis

d) None of these

Q227. Which one is green manure—————-

a) Sesbania b) Cyanobacteria

c) Azolla d) Anabaena

Q228. Azolla is used as biofertilizer it contains————-

a) Cyanobacteria b) BGA

c) Algae d) Fern

Q229. The most quickly available source of nitrogen to plants are-

a) Nitrate fertilizers
b) Nitrite fertilizers
c) Nitrite fertilizers
d) None of these

Q230. Most effective pesticides are-—————-

a) Organophosphates
b) Organometals
c) Dyes
d) None of these

Q231. Which is true for DDT—————

a) A biodegradable pesticiceds
b) A non- biodegradable pesticicde
c) Biopesticide
d) None of these

Q232. A major componenet of Baurdex mixture is————

a) Coppper sulphate
b) Ferrous sulphate
c) Copper oxide
d) None of these

Q233. Which one is organochloride————

a) Endosulphan
b) Endosulphate
c) Endocarbonate
d) Endobororon

Q234. IPM stands for—————

a) Integrated Pest Management
b) Integrated pot management
c) Both a & b
d) None of these

Q235. Green manure increases crop yield by—————

a) 30-50%
b) 10-70%
c) 5-15%
d) 8-80%

Q236. Insecticicdes genrally affects——————-

a) Respiratory system
b) Circulatory system
c) Nervous system
d) Digestive system

Q237. Organism associated with sorghum and cotton which provide nitrogen to them are

a) *Azospirrilum* & *Azotobacter*

b) *Spirullina* & *Azotobacter*

c) *Azotobacter* alone

d) None of these

Q238. Which one of these is not used as Bioperticide

a) *Bacillus thuringenesis* b) *Pseudomonas putida*

c) *Trichoderma* spp d) *Xanthomonas compestris*

Q239. The process of preserving food by rapid freezing followed by dehydration under vaccum is called

a) Lyophilization b) Thawing

c) Storage at 4^0C d) None of these

Q240. Cold sterilization refers to the preservation of food by——

a) Radiation b) Chemicals

c) Gases d) Fumes

Q241. Main purpose of blanching during canning is————

a) To soften products

b) To denature enzymes

c) To reduce microbial populations

d) All

Q242. Which of the method dehydrates microbial cells by plasmolysis thereby killing them

a) Sugaring b) Smoking

c) Heating d) Pasteurization

Q243. The process of preserving meat by storing in a covered earthenware jug is called

a) Burial b) Curing

c) Jugging d) All of these

Q244. Salt is added to meat for preservation so that it absorbs

a) Moisture b) Air

c) Softness d) All of these

Q245. Undesirable change in food that make it unsafe for human consumption is called as-

a) Food spoilage b) Food contamination

c) Food preservation d) All of these

Q246. Food preservation involves——————

a) To incresase shelf life of food

b) To ensure safety for human consumption

c) Both a & b

d) None of these

Q247. Common food poisoning microbes are——————

a) *Clostridium* b) *Salmonella*

c) Both a & b d) None of these

Q248. Botulinum is caused by——————

a) *C. botulinum* b) *Bacillus botulinum*

c) *Salmonella* d) All of these

Q249. Which of the following is true regarding botulinum toxin—

a) Produce by *C. botulinum* b) Gram posistive bacteria

c) Anaerobic Bacteria d) All of these

Q250. *Clostridium perfingens* poisioning is associated with——

a) Meat products b) Milk products

c) Beverages d) All of these

Q251. Major carrier of Salmonellosis are——————-

a) Meat and Egg b) Milk product

c) Poultry d) All of these

Q252. Aflotoxin is produced by————-

a) Aspergillus species b) Bacteria

c) Penicillium d) None of these

Q253. Use of living organism to degrade environmental pollution is called

a) Microremediation b) Bioremediation

c) Nanoremediation d) All

Q254. Which of the following bacteria is called superbug———

a) *Bacillus subtilis* b) *Pseudomonas putida*

c) *Trichoderma* d) None of these

Q255. Process of extracting metals from ore is called————-

a) Bioleaching b) Bioextraction

c) Biofilteration d) None of these

Q256. Process of converting environmental pollutants into harmless byproducts by naturally occurring microbes is called

a) Ex-situ bioremediation b) In-situ bioremediation

c) Extrensic bioremediation d) Intrinsic bioremediation

Q257. Chlorella species are widely used for the removal of——

a) Heavy metals b) Organic Wastes

c) Hydrocarbons d) All of these

Q258. A non- directed physicochemical interaction between heavy metal ion and microbial surface is called

a) Biotransformation b) Bioconversion

c) Biosorption d) Biomining

Q259. 2,4-D is an effective————

a) Weedicide b) Herbecide

c) Biopesticide d) Insecticide

Q260. Which of the following pesticide is acetylcholine esterase inhibitor——

a) BHC
b) Endosulphan
c) Malathion
d) Aldrin

Q261. What is agent orange————-

a) A weedicide containing dioxin
b) A herbiicide containing dioxin
c) A weedicide
d) A herbicide

Q262. A good example of biofertilizer which improves P uptake is————

a) Azospirillum
b) Rhizobium
c) Actinomycetes
d) All of these

Q263. A major ingradient of penicillin production media is—-

a) Corn meal
b) Corn steep liquor
c) Cane steep liquor
d) None of these

Q264. In the industrial production of streptomycin, the secondary metabolite or byproducts is

a) Vitamin – B12
b) Vitamin – C
c) Vitamin – B6
d) Ethanol

Q265. The pH, to be maintained for the production of penicillin is————

a) 7.5
b) 6.5
c) 8.0
d) 5.0

Q266. Citric acid is used as——————————-

a) Flavouring agent in food
b) As an antioxident
c) As preservative
d) All of the above

Q267. Citric acid is produced in aerobic conditions by the fungi —
____________.

a) *Aspergillus* b) Penicillin

c) *Mucor* d) All of these

Q268. The raw material for citric acid production is —

a) Corn b) Molasses

c) Starch d) None of these

Q269. *Aspergillus niger* is used generally for the production of —
____

a) Ethanol b) Penicillin

c) Citric acid d) Lactic acid

Q270. The by-product during streptomycin production is ————
————.

a) Vitamin A b) Proline

c) Vitamin B12 d) None of these

Q271. Enzymes, antibodies, hormones and hemoglobin are examples of——

a) Fibrous proteins b) Glandular proteins

c) Globular proteins d) Enzymatic proteins

ANSWERS

1.	c	2.	c	3.	d	4.	d	5.	a
6.	a	7.	c	8.	a	9.	a	10.	c
11.	d	12.	a	13.	b	14.	d	15.	c
16.	b	17.	d	18.	b	19.	d	20.	a
21.	c	22.	a	23.	a	24.	b	25.	a
26.	b	27.	b	28.	b	29.	a	30.	d
31.	a	32.	d	33.	c	34.	b	35.	b
36.	a	37.	b	38.	b	39.	a	40.	c
41.	b	42.	d	43.	c	44.	d	45.	c

46.	a	47.	c	48.	b	49.	d	50.	c
51.	c	52.	a	53.	a	54.	b	55.	d
56.	b	57.	c	58.	a	59.	d	60.	a
61.	a	62.	b	63.	a	64.	a	65.	a
66.	d	67.	d	68.	d	69.	c	70.	b
71.	d	72.	a	73.	a	74.	b	75.	b
76.	d	77.	b	78.	b	79.	b	80.	a
81.	b	82.	b	83.	a	84.	b	85.	c
86.	c	87.	d	88.	c	89.	a	90.	d
91.	d	92.	b	93.	c	94.	a	95.	a
96.	a	97.	b	98.	d	99.	c	100.	b
101.	b	102.	a	103.	d	104.	d	105.	b
106.	a	107.	c	108.	a	109.	c	110.	d
111.	a	112.	d	113.	c	114.	a	115.	d
116.	c	117.	a	118.	a	119.	a	120.	b
121.	c	122.	c	123.	d	124.	c	125.	a
126.	a	127.	a	128.	a	129.	d	130.	d
131.	a	132.	a	133.	a	134.	b	135.	a
136.	a	137.	a	138.	b	139.	a	140.	a
141.	d	142.	a	143.	b	144.	b	145.	b
146.	c	147.	a	148.	b	149.	a	150.	d
151.	d	152.	d	153.	c	154.	a	155.	a
156.	a	157.	c	158.	b	159.	b	160.	b
161.	a	162.	d	163.	b	164.	d	165.	d
166.	b	167.	b	168.	b	169.	b	170.	a
171.	a	172.	a	173.	c	174.	a	175.	a
176.	a	177.	a	178.	a	179.	a	180.	c
181.	a	182.	a	183.	a	184.	a	185.	a
186.	a	187.	a	188.	a	189.	a	190.	c
191.	a	192.	c	193.	a	194.	a	195.	a
196.	a	197.	c	198.	a	199.	b	200.	a
201.	c	202.	c	203.	a	204.	a	205.	a
206.	a	207.	a	208.	a	209.	a	210.	b

211. b	212. a	213. a	214. a	215. a
216. a	217. a	218. a	219. a	220. a
221. a	222. c	223. a	224. a	225. a
226. a	227. a	228. a	229. a	230. a
231. b	232. a	233. a	234. a	235. a
236. a	237. a	238. d	239. a	240. a
241. d	242. a	243. c	244. a	245. a
246. c	247. c	248. a	249. d	250. a
251. a	252. a	253. b	254. b	255. a
256. d	257. a	258. b	259. a	260. c
261. a	262. a	263. b	264. a	265. b
266. d	267. a	268. a	269. c	270. c
271. c				

Unit-VI

Microbial Genetics

Q1. A mutant bacterium requires the addition of some growth factor because they are not able to grow on a minimal medium is known as which one of the following?

a) Auxotrophs b) Culture media

c) Hetrotrophs d) None of these

Q2. The degradation of foreign genetic material by which one of the following enzymes after the genetic material enters a host cell is known as host restrictions?

a) Nuclease b) Helicases

c) DNA Ligase d) None of the above

Q3. First time Joshua Lederberg and Edward L. Tatum (1946) presented experiment for the evidence of?

a) Bacterial conjugation b) Transduction

c) Transformation d) None of these

Q4. The word used for the small solid support onto which are spotted hundreads of thousands of tiny drops of DNA that canbe used to screen gene expression is

a) Southern blot b) cloning library

c) DNA microarrays d) northern blot

Q5. Special appendages are present on the bacterial surface which forms the conjugation tube?

a) *F. pili* b) Fimbriae

c) Flgaella d) Mesosomes

Q6. The transfer of genetic material from one cell to another by a bacteriophage is called the?

a) Conjugation b) Transformation

c) Transduction d) None of these

Q7. The phenomenon of transduction was first discovered by ?

a) Huge-de vries b) Griffth

c) Bernard d) Zinder and Lederberg

Q8. What is virulent Phage?

a) The phage that reproduce by using a lytic cycle

b) The phage that reproduce by using lysis cycle

c) Both A and B

d) None of the above

Q9. What is the function of topoisomerase IV?

a) Unlinking the DNA double helix

b) Unwinds the DNA double helix

c) Proof reading

d) None of the above

Q10. The zone of unwound DNA where replication occurs is known as the?

a) Replication fork b) 3 end

c) 5 end d) None of these

Q11. The ability of a cell to take up DNA fragments from the surroundings is called as ——————

a) Fitness b) Fecundity

c) Competence d) Hfr

Q12. Which enzymes helps in the unwinding of DNA double helix and expose a short single strands region?

a) Poly 1 b) Topoisiomerase

c) Helicase d) Nuclease

Q13. The phenomenon of moving genetic segments from one locations to the other in a genome is known as a ?

a) Transduction b) Transition

c) Both A and B d) Transposition

Q14. Mc Clintock Barbara was Awarded Nobel prize in 1983 for consistent work on?

a) Transposition b) Mutation

c) Jumping gene d) Replication

Q15. What is abortive transduction?

a) When DNA is not integrated into the endogenote and remain free

b) When exogenote compeletely integrated into th recipient bacterial chromosomes

c) When exogenote may be integrated into th recipient bacterial chromosomes

d) None of these

Q16. Generalized transduction was discovered by the?

a) Lederberg and woolman b) Norton And Zinder

c) Iwanwosky d) None of the above

Q17. When latent of phage genome that remains within the host without harm and integrated with chromosomes is called?

a) prophage b) episome

c) Hfr d) none of these

Q18. Who used the term mutation for first time?

a) Huge-de vries b) Lederberg

c) *Bernard Davis* d) Mc Clintock Barbara

Q19. Which one of the following was the first bacterial genome to be sequenced?

a) Arabidopsis thaliana b) Bacteriophage MS2

c) Haemophilus influenzae d) Phage O-X174

Q20. The unit of mutation within the gene is known as which one of the following———-

a) Muton b) Intron

c) Recon d) Cistron

Q21. Which one of the following is known as the building block of the nucleic acids?

a) Amino acid b) NucleoProtien

c) Nucleosides d) Nucleotides

Q22. Which one of the following is fertility factor in bacterium?

a) F factor b) S factor

c) Both a and b d) None of the above

Q23. Which one of the following processes describe to the production of RNA from DNA?

a) RNA splicing b) Transcription

c) Translation d) Transposition

Q24. The RNA world was coined first time by which one of the following scientist?

a) Sidney Altman b) Thomas Cech

c) Carl woese d) Walter Gilbert

Q25. A set of nucleotide triplet that will code one amino acid is termed as which one of the following ?

a) Anticodon b) Cistron

c) Codon d) Muton

Q26. During the protein synthesis, which one of the following RNA acts as a carrier of amino acids?

a) mRNA b) rRNA

c) tRNA d) None of the above

Q27. The term genome refers to all the——————

a) DNA present in a cell b) Protein present in a cell

c) RNA is present in a cell d) None of the above

Q28. Flow of genetic information from one generation to other generation is called————: the flow genetic information with in a single cell a process called a ———

a) Replication and gene elongation

b) Replication and gene expression

c) Replication and transcription

d) Replication and translation

Q29. RNA polymerase enzymes catalyze the——————

a) Replication b) Translation

c) transcription d) none of the above

Q30. The DNA and RNA are the polymers of the nucleotides linked together by the————

a) Phosphodiester bond b) Sulphudes bomd

c) Hydrogen bond d) Ionic bond

Q31. The DNA organized in the form of a closed circle is—-

a) Archaea and mostly bacteria

b) Only archaea

c) Only eukaryotes

d) Only bacteria

Q32. The DNA replication is————

a) Semi-conservative

b) Conservative

c) Sometime semicinservative sometime conservative

d) All of these

Q33. What is the function of SSB Protien ?

a) Bind single strand DNA after strand separated by helicase

b) Termination of primers

c) Fills gaps in DNA Formed by removal of RNA primer

d) Helps in the termination at replication

Q34. Lysogenic conversion————

a) is the induction of a prophage to its virulent state

b) is the immunity that a prophage confers on a bacterium.

c) is a change in pathogenicity due to the presence of a prophage.

d) refers to the incorporation of a prophage into the chromosome.

Q35. Which one type of ring is related to a Purine?

a) Single ring b) Double ring

c) Triple ring d) None of the above

Q36. Operon are known for metabolisim of which of the following in *E. coli*, except?

a) Arginine

b) Tryptophan

c) Arabinose

d) Sucrose inducing enzyme

Q37. Which one of the following mutation causing a substitution of one amino acid?———

a) Missense mutation b) Point mutation

c) Silent mutation d) None of the above

Q38. In which of the following, RNA acts as agenetic material?

a) *E.coli* b) *Neurospora*

c) TMV virus d) None of the above

Q39. During protein synthesis, which one of the following RNA act as a carrier of amino acid?

a) mRNA b) rRNA

c) tRNA d) None of the above

Q40. In the helical strands of DNA, which one of the following is the distance between each turn?

a) 20 A b) 28 A

c) 34A d) 42A

Q41. The PCR reaction was developed by the ——

a) Kary Mullis b) Walter Gilbert

c) Thomas Cech d) Sidney Altman

Q42. Which of the example is not true about the Recombinant DNA molecule ?

a) Plasmid b) Phgae

c) Virus d) Mesosome

Q43. In RDT which is called a molecular scissors?

a) DNA ligase b) Vector

c) Restriction endonuclease d) Heilcase

Q44. When a suitable DNA molecule capable of self replication in the selected host is called

a) Plasmid b) Vector

c) Mesosome d) None of the above

Q45. Which is known as molecular glue in RDT?

a) Helicase b) Endonulease

c) Ligase d) Polymerase

Q46. Which of the following is used to remove the 5' – phosphate from the DNA

a) Linker b) Helicase

c) Alkaline phosphatase d) Exonuclease

Q47. Which of the following is used to add the phosphate group to an end having a free 5'OH ?

a) Terminal deoxynuleotidyl transferase

b) Lambda exonucllease

c) Reverse transcriptaese

d) T4 polynucleotide kinase

Q48. Endonuclease an enzyme that are produce a internal cut, called.... In DNA molecule?

a) Cleavage b) Point

c) Gap d) Nick

Q49. RNA dependent DNA polymerase discovered by————-

a) Carl Richard Woese b) Paul Ehrlich

c) Temin and Baltimore d) None of the above

Q50. Good properties of vector except?

a) It should be able to replicate autonomously

b) A vector should be ideally more than 20 kb in size

c) The vector should be easy to isolate and purify

d) The vector should contain a unique target sites

Q51. Puff are the sites of which one of the following synthesis?

a) DNA synthesis b) RNA synthesis

c) Protien synthesis d) None of the above

Q52. If a base replaces by another base pair, sequence mutation, resulting replacement is known as —————

a) Alternation b) Substitution

c) Missense d) Nonsense

Q53. Bacteria composed single chromosomes, having single copy of gene, known as———-

a) Diploid b) Haploid

c) Polyploidy d) Monoploid

Q54. Substitutions that prematurely stops synthesis of protein, by generating stop codon, called as —————

a) Missense mutation b) Nonsense mutation

c) Frameshift mutation d) Alternation

Q55. What are approximate numbers of base pairs in bacterial DNA ?

a) 5×10^{10}
b) 5×10^{6}
c) 5×10^{8}
d) 5×10^{11}

Q56. When one or more base pair are deleted or added in sequence, is called

a) Substitution mutation
b) Missense mutation
c) Nonsense mutation
d) Frameshift mutation

Q57. A gene is best defined as

a) A segment of DNA
b) Three nucleotides that code for an amino acid
c) A sequence of nucleotides in DNA that codes for a functional product.
d) A sequence of nucleotides in RNA that codes for a functional product.

Q58. Transcription in bacteria is performed by which of the following?

a) RNA-dependent RNA polymerase
b) RNA-dependent DNA polymerase
c) DNA-dependent RNA polymerase
d) DNA-dependent DNA polymerase

Q59. Many bacterial translation factors are G proteins with which of the following activities?

a) ATPase activity
b) GTPase activity
c) GDPase activity
d) GMPase activity

Q60. During the transcription of a certain protein, an extra cytosine was placed into a gene region, throwing off the correct amino acid sequence. What type of mutation occurred?

a) Transposon
b) Insertion
c) Base-pair substitution
d) Deletion

Q61. DNA requires which of the following?

a) Sunlight
c) Protein synthase
b) Meiotic division
d) Many enzymes

Q62. Why is there no duplication of the DNA between meiosis I and meiosis II?

a) To produce genetically identical daughter cells
b) To increase genetic variability
c) To reduce the chromosome number to haploid in the resulting daughter cells
d) The chromosomes duplicate twice during meiosis I.

Q63. The genetic disorder sickle-cell anemia is an example of—-

a) Pleiotropy.
b) Heterozygous dominance.
c) Homozygous dominance.
d) Epistasis.

Q64. What is the key difference between mitosis in plant cells and mitosis in animal cells?

a) The chromosomes aren't duplicated during interphase in plant cells.
b) Animal cell mitosis results in two daughter cells; plant mitosis produces three.
c) The two daughter cells formed in plant cell mitosis aren't genetically identical.
d) A cell plate is formed during mitosis in plant cells.

Q65. Which of the following genotypes causes Klinefelter syndrome?

a) XO
c) XXY
b) XX
d) XYY

Q66. Which one of the following fat soluble vitamin serves in gene regulation?

a) Vitamin A
b) Vitamin D
c) Vitamin E
d) Vitamin K

Q67. Why do researchers use DNA polymerase from a bacterium found in superheated water for PCR?

a) It's the only bacterium that contains this enzyme.
b) It can withstand the elevated temperatures that are required to unwind DNA.
c) DNA must go through high-temperature sterilization before PCR can occur.
d) These bacteria contain more DNA polymerase than any other species.

Q68. Bacteria composed single chromosomes, having single copy of gene?

a) Diploid
b) Haploid
c) Polyploidy
d) Monoploid

Q69. How much time is taken by the new DNA tomake a whole phgae particle?

a) 24 hours
b) 60 minutes
c) 2 minutes
d) 10-20 minutes

Q70. Which of the following strands of DNA would be the complement strand to C-C-A-T-C-G?

a) G-G-T-A-G-C
b) A-A-C-G-A-T
c) G-G-A-T-G-C
d) T-T-G-C-T-A

Q71. A recombinant cell————

a) is a cell that receives DNA from an outside source and incorporates it into its own.
b) A bacterial plasmid
c) the cells must come into contact with each other.
d) None of the above

Q72. What is the range of minimum set of genes required for life?

a) 50-100 genes b) 250-350 genes

c) 1000-1500 genes d) 1500-2000 genes

Q73. The flow of genetic material in microbial cells usually takes place from

a) RNA through DNA to proteins

b) Proteins through RNA to DNA

c) DNA through RNA to proteins

d) None of these

Q74. Which of the following is used for determining the location of specific genes within the genome?

a) Genomics b) Annotation

c) Cloning d) Proteomics

Q75. Proteomics is ————

a) The study of algal genomes

b) A branch of quantum physics dealing with proteins

c) The study of formation of lipo-protein in animals

d) The study of the entire collection of proteins expressed by an organism

Q76. Which of the following is concerned with the management and analysis of biological data using computers—-

a) Bio-chem b) Bio-physics

c) Genomics d) Biomechanics

Q77. If you wanted to produce a certain protein in the laboratory, which of the following would be necessary for establishing the correct amino acid sequence?

a) Ribosomal proteins c) rRNA

b) mRNA | d) Anticodons

Q78. Why Deinococcus radiodurans is able to survive massive exposure to radiation?

a) Because it produces a thick shell which acts as a shield from the radiation

b) Because it has unique DNA repair mechanisms

c) Because its cellwall contains radioactive elements

d) Because it has many copies of genes encoding DNA repair

Q79. In the time since E. coli and Salmonella diverged evolutionarily————

a) There has been little change in either genome

b) E. coli has acquired many genes via horizontal transfer

c) E. coli has lost approximately 50% of its genome

d) None of these

Q80. The physical nature of genomes is studied under———

a) Structural genomics

b) Comparative genomics

c) Proteo genomics

d) Functional genomics

Q81. The species of bacteria, which possesses 250 genes for lipid biosynthesis is

a) *M. genitalium* b) *M. tuberculosis*

c) *E. coli* d) *H. influenzae*

Q82. Why the bacterium Treponema pallidum is difficult to culture?

a) Because it lacks the genes needed for TCA cycle and oxidative phosphorylation

b) Because it requires extremely low temperature at which water freezes

c) Because it is unable to use carbohydrates as an energy source

d) Because it requires a great deal of water to reproduce

Q83. Tautomerization is an example of spontaneous mutation—-

a) True b) False

Q84. Which of the following statements concerning recombinant DNA technology is false?

a) Mutant strains of bacteria produced by genetic recombination are often unable to survive in the natural environment.

b) Recombinant DNA technology provides a high degree of risk to the health of the general population.

c) Production of large amounts of proteins such as insulin and human growth hormone has been made possible using recombinant DNA technology.

d) Recombinant DNA technology offers specific benefits to the scientific, medical, and general population.

Q85. Which of the following events is most likely due to bacterial transformation

a) A non-encapsulated strain of Streptococcus pneumoniae acquires a gene for capsule formation from the extract of an encapsulated strain.

b) A gene for gentamicin resistance from an Escherichia coli chromosome

c) A strain of Neisseria gonorrhoeae starts producing a plasmid-encoded beta-lactamase similar to that another Gram-negative strain

d) A formerly non-toxigenic strain of Corynebacterium diphtheriae becomes toxigenic

Q86. The first vaccine for human use produced using recombinant DNA technology was the:

a) AIDS vaccine b) Hepatitis A vaccine

c) Hepatitis B vaccine d) Polio vaccine

Q87. Each of the following events involves recombinantion of DNA EXCEPT-

a) Transduction of a chromosomal gene

b) Transposition of a mobile genetic element

c) Integration of a temperate bacteriophage

d) Conjugation, eg, transfer of an R (resistance) factor

Q88. Which type of genetic exchange in bacteria is susceptible to the activity of deoxyribonuclease?

a) Transfection b) Conjugation

c) transduction d) Transformation

Q89. The form of genetic exchange in which donor DNA is introduced to the recipient by a bacterial virus is

a) Transduction b) Horizontal transfer

c) Conjugation d) Transfection

Q90. The ability of a cell to bind DNA to its surface and import it is required for which genetic process ——

a) Specialized transduction

b) Conjugation

c) Homologous recombination

d) Generalized transduction

Q91. The nucleosome is ——————

a) Found in nucleus

b) Composed of RNA

c) Composed of DNA and proteins

d) Found in cell way

Q92. A protein containing partial with no detachable 'nucleic acid' are known as ——————-

a) Virion b) Prions

c) Viroid d) Proteins

Q93. Integration of viral 'DNA' into cell 'DNA' results in a structure named as

a) Viral genome b) Prophage

c) Virion d) Prion

Q94. A bactericidal drug that inhibits activity of DNA gyrase, ultimately inhibit synthesis of————

a) Bacterial RNA b) Bacterial DNA

c) Ribosome d) Mitochondria

Q95. Viruses that are defective, having proteins and nucleic acid, but, cannot replicate without————

a) Bacteriophage b) Virion

c) Prion d) Helper virus

Q96. Papovavirus' is the————

a) Tumor virus b) RNA Tumor virus

c) Enveloped viruses d) Naked viruses

Q97. After completion of a viral replication cycle number of progeny in host cell is, approximately——

a) 100 virion b) 200 virion

c) 50 virion d) 40 virion

Q98. 'Viroid' cause diseases rarely in humans, but , mostly in— ———-

a) Plants b) Animals

c) Horses d) Camels

Q99. Genetic material of bacteria is composed of, a double-stranded ——-

a) Linear DNA
b) Ladder like DNA
c) Coiled DNA
d) Circular DNA

Q100. Transduction was discovered by————

a) Griffith
b) Zinder and Lederberg
c) Lederberg, Hayes and Woolman
d) Iwanowsky

Q101. If the F factor is attached to the bacterial genome the donor is called as————

a) F +
b) F++
c) F+ super strain
d) Hfr strains

Q102. The ability of a cell to take up DNA fragments from the surroundings is called as————-

a) Fitness
b) Fecundity
c) Competence
d) Hfr

Q103. The unidirectional transfer of genetic material from a donor bacterium to recipient bacterium by cell to cell contact is termed as

a) Transformation
b) Conjugation
c) Recombination
d) Transduction

Q104. The transfer of genetic material from one bacterium to another via virus is called————-

a) Transformation
b) Conjugation
c) Recombination
d) Transduction

Q105. The uptake of DNA fragments from surroundings by a bacterium is termed as————

a) Transformation
b) Conjugation
c) Recombination
d) Transduction

Q106. Bacteria reproduce vegetatively by———

a) Fission only
b) Fission and fragmentation
c) Fission, fragmentation and budding
d) None of the above

Q107. Bacteria reproduce asexually by———

a) Conjugation b) Amitosis
c) Meiosis d) Transformation

Q108. Binary fission in bacteria involves all except———

a) Cell elongation b) Cytokinesis
c) DNA duplication d) Spindle formation

Q109 . Minute bodies that are formed during fragmentation in bacteria are called———

a) Conidia b) Gonidia
c) Microconidia d) Oospore

Q110. Endospore formation is a method to tide over unfavorable condition which is generally seen in———

a) Gram positive bacteria
b) Gram negative bacteria
c) Gram positive bacteria and Gram negative bacteria
d) Streptomycetes

Q111. Which of the following favors endospore formation ———

a) High concentration of nitrogenous nutrients
b) Low concentration of nitrogenous nutrients
c) Low concentration of oxygen
d) None of these

Q112. The partial diploids formed as a result of sexual reproduction in bacteria is termed as———

a) Zygotes b) Hetero

c) Prozygote d) Merozygote

Q113. The Physical nature of genomes is studied under?

a) Structural genomics b) Comparative genomics

c) Proteo genomics d) Fucntional genomics

Q114. In a protein synthesis , a unit of function by which of the following that specific one polypeptide chain in DNA is?

a) Cistron b) Muton

c) Recon d) None of the following

Q115. Which one of the following scientist by proposed "Operon model"

a) Britten b) Francois Jacob

c) Jacques Monod d) Both A and C

Q116. A ―――――――― is produced by joining together two or more DNA segments usually originating from different organisms――――――――

a) Competent cell

b) Plasmid

c) Vector

d) Recombinant DNA molecule

ANSWERS

1.	a	2.	a	3.	a	4.	c	5.	a
6.	c	7.	d	8.	c	9.	a	10.	a
11.	c	12.	c	13.	d	14.	c	15.	a
16.	b	17	a	18.	a	19.	c	20.	a
21.	d	22.	a	23.	b	24.	d	25.	c
26.	c	27.	a	28.	b	29.	c	30.	a
31.	a	32.	b	33.	a	34.	d	35.	b

36. a	37. b	38. c	39. c	40. c
41. a	42. d	43. c	44. a	45. c
46. c	47. d	48. a	49. c	50. b
51. a	52. b	53. b	54. a	55. b
56. d	57. c	58. a	59. b	60. b
61. d	62. c	63. a	64. d	65. c
66. b	67. b	68. b	69. d	70. a
71. a	72. b	73. c	74. b	75. d
76. b	77. b	78. d	79. c	80. a
81. b	82. a	83. b	84. b	85. a
86. c	87. d	88. d	89. a	90. d
91. c	92. b	93. b	94. b	95. d
96. a	97. a	98. a	99. d	100. b
101. d	102. c	103. b	104. d	105. a
106. c	107. b	108. d	109. b	110. a
111. b	112. d	113. a	114. a	115. d
116. d				

Miscellaneous Questions

Q1. Fusion between a plasma cell and a tumor cell creates a

a) Myeloma
b) Natural killer cell
c) Lymphoblast
d) Lymphoma
e) Hybridoma

Q2. B cells that produce and release large amounts of antibody are called:

a) Memory cells
b) Basophils
c) Plasma cells
d) Killer cells
e) Neutrophils

Q3. Which of the following immune cells/molecules are most effective at destroying intracellular pathogens?

a) T helper cells
b) B cells
c) Antibodies
d) Complement
e) T cytolytic cells

Q4. A living microbe with reduced virulence that is used for vaccination is considered:

a) A toxoid
b) Dormant
c) Virulent
d) Attenuated
e) Denatured

Q5. Which of the following convey the longest lasting immunity to an infectious agent?

a) Naturally acquired passive immunity
b) Artificially acquired passive immunity

c) Naturally acquired active immunity

d) All of these

e) None of these

Q6. Which of the following substances will not stimulate an immune response unless they are bound to a larger molecule?

a) Antigen b) Virus

c) Hapten d) Miligen

e) Antibody

Q7. How much of globulin is present in human serum?

a) 8% b) 12%

c) 16% d) 4%

Q8. The substance which acts as antimetabolites are called —-

a) Activators b) Substrates

c) Inhibitor d) Cofactor

Q9. Enzymes are chemically ————————-

a) Lipids b) Proteins

c) Carbohydrates d) None of these

Q10. In a antigen haptens are ————————

a) Immunogenic b) Non-immunogenic

c) Antigenic d) None of these

Q11. The antibody that is first formed after infection is ———

a) IgG b) IgM

c) IgD d) IgE

Q12. Antibodies in our body are produced by —————

a) B-lymphocytes b) T-lymphocytes

c) Monocytes d) RBC's

Q13. Syphillis is caused by ————————-

a) *Staphylococcus aureuss* b) *Yersinia putis*

c) *Treponema pallidum* d) *Streptococcus syphilitic*

Q14. The use of monoclonal antibodies is ————-

a) Immunotherapy b) Gene therapy

c) Blood transfusion d) Organ transfusion

Q15. Hybridoma technique is used for ————

a) Monoclonal antibodies b) Polyclonal antibodies

c) Both a and b d) None of these

Q16. Test used for AIDS is————

a) Widal test b) ELISA

c) Aggluatination d) CFT

Q17. Which of the following can provide naturally acquired passive immunity for the new born.

a) IgA b) IgG

c) IgE d) IgM

Q18. AIDS disease is caused by a virus which belongs to ——

a) Retro virus group b) Rhabdo virus group

c) Hepatitis virus group d) Adeno virus group

Q19. Pasteur developed the vaccines for ————

a) Anthrax b) Rabies

c) Chicken cholera d) All of the above

Q20. Delayed type of hypersensitivity is seen in ————-

a) Penicillin allergy b) Contact dermatitis

c) Arthus reaction d) Anaphylaxis

Q21. The primary cells involved in immune response are————

a) NK-cells b) K-cells

c) Lymphocytes d) None of these

Q22. Antibodies are ————

a) Proteins b) Glycoproteins

c) Phospholipids d) None of these

Q23. General purpose antibody is ——————

a) IgA
b) IgG
c) IgM
d) IgD

Q24. Antibody present in colostrums is ——-

a) IgG
b) IgA
c) IgM
d) IgE

Q25. Which antibody is called millionaire molecule? -

a) IgA
b) IgM
c) IgG
d) IgD

Q26. Shick test is used for the detection of——————

a) Diphtheria
b) T.B.
c) Cholera
d) Typhoid

Q27. Pertussis vaccine is ——————

a) Heat killed
b) Formalin killed
c) Attenuated
d) live

Q28. DPT is ——————

a) Triple vaccine
b) Double vaccine
c) Tetanus toxoid
d) All of these

Q29. Vaccines are prepared from killed microbes, they are —-

a) Inactivated (killed) vaccine
b) Attenuated vaccines
c) Autogenous vaccine
d) None of these

Q30. Vaccines used against viral infections are ——————-

a) Measles and Mumps vaccine
b) Cholera vaccine
c) Typhoid vaccine
d) Anti-rickettsial vaccine

Q31. Vaccines prepared from toxins and chemicals are ——-

a) Cellular vaccines	b) Sub-cellular vaccines

c) Attenuated vaccines	d) Heterologous vaccines

Q32. Example for live vaccine is ——————-

a) Rubella & BCG	b) Polio & TAB

c) Diphtheria & Tetanus	d) Hepatitis A & Rabies

Q33. DPT is given for the prevention of ——————-

a) Diphtheria, Tetanus

b) Diphtheria, Pertusis

c) Diphtheria, Tetanus & pertusis

d) None of these

Q34. Antigenic variation is most extensive in ——————

a) Influenza virus	b) Small pox virus

c) Measles virus	d) Herpes virus

Q35. Which is the correct statement related to hepatitis B virus? ————

a) Paramyxo virus	b) Orthomyxo virus

c) Reo viruses	d) Retro viruses

Q36. Koplic's spots will develop in ——————

a) HIV	b) Measles

c) Mumps	d) Rubella

Q37. Viral DNA is resistant to DNA of the host cell because it contains

a) 5'-HMC	b) 5'-HMA

c) 5'-CHM	d) 5'MHC

Q38. Which of the following is an example of live vaccine?

a) Pertusis	b) Mumps

c) Cholera	d) Rabies

Q39. Triple toxoid vaccine gives protection against ———

a) Diphtheria, tetanus and rabies

b) Tetanus, whooping cough, Tuberculosis

c) Whooping cough, tetanus and Diphtheria

d) Whooping cough, cancer and T.B.

Q40. AIDS is caused by ————————

a) Retrovirus b) Prion

c) Rhabdovirus d) Retroprison

Q41. Penicillin is a ————————

a) Primary metabolite b) Secondary metabolite

c) Tertiary metabolite d) None of the above

Q42. The rejection of an organ transplant such as a kidney transplant, is an example of_____ Hypersensitivity.

a) Immediate b) Delayed

c) Allergy d) None of these

Q43. Listeriosis was ______ disease————————-

a) Food borne b) Water borne

c) Milk borne d) Air borne

Q44. Pus-forming forms are called as ————————

a) Pyoderm b) Pyogenic

c) Pyrogen d) None of the above

Q45. In Elisa technique, the antibodies are labeled by ———-

a) Acridine orange b) Alkaline phosphate

c) Neutral red d) Bromothymol blue

Q46. Botulism means ——————————-

a) Food adultration

b) Food poisioning by streptococcus bacteria

c) Chemical contamination of food

d) Food processing

Q47. Chloramphenicol is obtained from ——————

a) *Streptomyces griseus* b) *Streptomyces venezuelae*

c) *Streptomyces pyrogenes* d) None of these

Q48. Streptomycin is obtained from ——————

a) *Streptococcus* species b) *Streptomyces griseus*

c) *Straphylococcus aureus* d) None of these

Q49. All of the following are bacteriostatic chemotherapeutic agents except

a) Bacitracin b) Chloramphenicol

c) Novobiocin d) Tetracycline

Q50. Beta lactum ring is present in ——————

a) Erythromycin b) Penicillin

c) Tetracyclins d) Chromphenical

Q51. Ciprofloxacin acts by inhibiting——————

a) Cellwall synthesis b) RNA synthesis

c) Folate synthesis d) DNA gyrase

Q52. Lyme disease is caused by ——————

a) Bacteria b) Fungi

c) Spirochaete d) Virus

Q53. Toxic shock syndrome is caused by ——————-

a) *Staph.albus* b) *Staph.aureus*

c) *Strep.viridana* d) None of these

Q54. Black water fever is caused by ——————-

a) *P.vivax* b) *P.falciparum*

c) *P.ovale* d) None of these

Q55. Mantoux test detects ——————

a) *M.tuberculosis* b) *Cynaobacteria*

c) *Clostridia* d) Both a and b

Q56. The antibiotic acting on cell wall is ————-

a) Bactracin b) Penicillin

c) Cyclosporine d) All of these

Q57. Aflatoxin is produced by ————-

a) *Aspergillus* sps b) *Penicillium* sps

c) *Alternaria* sps d) None of these

Q58. Penicillin is discovered by ————-

a) Fleming b) Pasteur

c) Koch d) None of these

Q59. Antibiotics used in combination may demonstrate ————

a) Synergism b) Antaginism

c) both d) None of these

Q60. Common cold is caused by ————

a) Adeno virus b) Corono virus

c) Hepatitis virus d) Pox virus

Q61. The causative agent of conjunctivitis ————-

a) Adeno virus b) Corono virus

c) Paramyxo virus d) None of these

Q62. Diphtheria is an example of ————-

a) Bacteraemia b) Pyaemia

c) Septicemia d) Toxaemia

Q63. Influenza is belonging to ————

a) Orthomyxoviridae b) Retroviridae

c) Both a and b d) None of these

Q64. The causative agent of tetanus is ————

a) *Clostridium botulinum* b) *Cl.tetani*

c) *Cl.welchii* d) *Cl.Perfringens*

Q65. Dengue virus is transmitted from man to man by the ——

a) Sand fly b) Ticks

c) Aedes aegypti d) Culex

Q66. Yellow fever is caused by——————

a) Bunya virus b) Calci virus

c) Arbo virus d) None of these

Q67. Vector for leishmaniasis is——————

a) Tick b) Mite

c) Sand fly d) Tsetse fly

Q68. Japanese encephalitis is caused by——————

a) Toga Viruses b) Arbo Viruses

c) Para myxo Viruses d) Ortho myxo Viruses

Q69. The antibiotic acting on cell wall is——————-

a) Penicillin b) Bacitracin

c) Cyclosporin d) All of the above

Q70. Nalidixic acid activity is due to ——————

a) The inhibition of DNA synthesis

b) Inhibition of protein synthesis

c) The inhibition of cell wall synthesis

d) Both B and C

Q71. The role that human play in the plague life cycle is——

a) Secondary reservoir

b) Primary transmission vector

c) Primary host

d) Accidental intruder in rat flea cycle

e) None of these

Q72. *E.coli* produce which type of toxins?

a) Exotoxins b) Endotoxins

c) Leucocidin d) Both a and b

Q73. Main causative organism of gas gangrene is————-

a) *Banthrax* b) *Clostridium tetani*

c) *Cl.deficile* d) *Cl.perfringens*

ANSWERS

1.	e	2.	c	3.	e	4.	d	5.	c
6.	c	7.	a	8.	c	9.	b	10.	b
11.	b	12.	b	13.	c	14.	a	15.	a
16.	b	17.	b	18.	a	19.	d	20.	b
21.	c	22.	b	23.	b	24.	b	25.	b
26.	a	27.	b	28.	a	29.	a	30.	a
31.	b	32.	a	33.	c	34.	a	35.	c
36.	b	37.	a	38.	c	39.	c	40.	a
41.	b	42.	a	43.	a	44.	b	45.	b
46.	c	47.	b	48.	a	49.	a	50.	b
51.	d	52.	c	53.	b	54.	b	55.	a
56.	d	57.	a	58.	a	59.	c	60.	b
61.	a	62.	d	63.	a	64.	b	65.	c
66.	c	67.	c	68.	b	69.	d	70.	a
71.	d	72.	b	73.	d				

General Agriculture One Line Questions

Q1. India's Rank in world in wheat production

Ans. Second

Q2. India's rank in Rice production

Ans. Second

Q3. India's Rank in world in total cereals production

Ans. Third

Q4. India's Rank in world in vegetable production

Ans. Second

Q5. India's Rank in world in fruits production

Ans. Second

Q6. India's Rank in world in total pulse production

Ans First

Q7. India's Rank in world in sugarcane production

Ans. Second

Q8. India's Rank in world in Tea production

Ans. Second

Q9. India's Rank in world in urea production

Ans. Seocnd

Q10. India's consumption of kilo cal/capita/day in gms

Ans. 55.7gm/day

Q11. India's Rank in world in DAP production

Ans. Third

Q12. India's Rank in world in milk production

Ans. First

Q13. State soil having highest % of zinc deficiency in India

Ans. Maharashtra

Q14. Highest Bio fertilizers production state

Ans. Tamil nadu

Q15. How many nutrients are included under nutrient based subsidy

Ans. 4 (N,P,K,S)

Q16. Nutrient based subsidy start in which year

Ans. 2011

Q17. In which year MSP was announced by Govt.

Ans. 1970

Q18. Which committee recommend the MSP

Ans. CASP

Q19. Which committee approve the MSP

Ans. Cabinet committee on Economic affairs

Q20. No. of oil seeds covered under a MSP

Ans. 7

Q21. No. of cereals covered under a MSP

Ans. 7

Q22. No. of fiber crops covered under a MSP

Ans. 2

Q23. No. of pulses covered under a MSP

Ans. 5

Q24. Total no. of crops covered under MSP regime

Ans. 26

Q25. Duartion of pre monsoon In india

Ans. March- may

Q26. Duartion of post monsoon

Ans. Oct- Dec

Q27. Average lowest rainfall occurs in

Ans. Jaislmer (Rajasthan)

Q28. Hanumantha Rao committee is associate with

Ans. Fertilizers

Q29. History of indain agriculture was written by

Ans. M.S Randhawa

Q30. First agriculture university

Ans. G.B Pant agri. University Pantnagar

Q31. State which have highest acidic soils

Ans. West Bengal

Q32. State which have highest saline soils

Ans. Gujarat

Q33. State which have highest use of pesticides

Ans. Punjab

Q34. The Indian journal of agriculture sciences published by

Ans. ICAR

Q35. Indian farming journal published by

Ans. ICAR

Q36. King Baudouin international development prize was given for

Ans. Green revolution

Q37. Dr. G.S Khush awarded by wolf prize for which crop

Ans. Rice

Q38. India's first NRC crop was

Ans. Groundnut

Q39. India's first all India Co- ordinate research project was for the crop

Ans. Maize

Q40. First agriculture minister of independent was

Ans. Rafi Ahmed kidwai

Q41. Present Director general of ICAR

Ans. Dr. Trilochan Mohpatra

Q42. The department of agriculture research and education (DARE) was established in the ministry of agriculture in

Ans. Dec 1973

Q43. Present secretary department of agricultural research and education (DARE)

Ans. Dr. Trilochan Mohpatra

Q44. Constitution of NCF (National commission of Farmers) in

Ans. 18 nov. 2004

Q45. Set up of Indian Tea board in year

Ans. 1st april 1954

Q46. Present director of IARI

Ans. Dr. Ravindra kaur

Q47. Black revolution is related to a

Ans. Biofuel

Q48. Year known as international rice year

Ans. 2004

Q49. year known as international forest year

Ans. 2011

Q50. Year known as international horticultural year

Ans. 2012

Q51. World food day is celebrate on

Ans. 16th oct

Q52. First soi ltesting lab was starting in year

Ans. 1955-56 (IARI)

Q53. Production of off season vegetable is known as

Ans. Vegetable forcing

Q54. Present president of ICAR

Ans. Union ministerof agriculture Dr. radha mohan singh

Q55. Founder chairman of NABARD

Ans. M. Ramakrishnayy

Q56. Cash prize given in ICAR N. Borlaug award

Ans. 10 lacs

Q57. Father of agronomy

Ans. Peter Dearsenzi

Q58. Father of weed sicence.

Ans. Jethero tull

Q56. Objective of sustainable of agriculture is

Ans. Ecological balance

Q57. A system of growing same crop on the same land year after year is known as

Ans. Mono cropping

Q58. Scientific Name of rice

Ans. *Oryza sativa* 2n= 24

Q59. Botanical name of wheat is

Ans. *Triticum aestivum*

Q60. Botanical name of maize

Ans. *Zea maysL.* Chromosome number (2n= 20)

Q61. Boatanical name of sorghum/ jowar

Ans. *Sorghum bicolor*

Q62. Scientific name of barley

Ans. *Hordeum vulgare*

Q63. Botanical name of Pearl Millet

Ans. *Pennistum glaucm* L.

Q64. Boatnical name of Gram

Ans. *Cicer arietinum* (Brown gram), *Cicer kabulianum* (kabuli white gram)

Q65. Boatanical name of arhar is

Ans *Cajanus cajan*

Q66. Botanical name of sugarcane

Ans *Sccharum offiicinarum*

Q67. Botanical name of ground nut

Ans. *Arachis hypogeal*

Q68. Botanical name of sunflower is

Ans *Helianthus annus*

Q69. Botanical name of soyabean

Ans. *Glycine max*

Q70. Boatanical name of oilseed

Ans *Limum Usitatisium*

Q71. Botanical name of Mustard

Ans. *Brassica* spp.

Q72. Botanical name of tomato is

Ans *Solanum lycopersicum*

Q73. Botanical name of potato is

Ans *Solanum tuberosum*

Q74. Botanical name of cauliflower

Ans. *Brassica oleracea* var botrytis

Q75. Botanical name of cabbage

Ans. *Brassica oleracea* var capitata

Q76. Boatanical name of mango

Ans. *Mangifera Indica*

Q78. Botanical name of banana

Ans. *Musa acuminate*

Q79. Botanical name of apple

Ans. *Malus domestica*

Q80. Black soil found in

Ans. Maharastra

Q81. Red soil found in

Ans. Tamil nadu

Q82. Humic acid soluble in

Ans. Alkali solution

Q83. C: N ratio of humus

Ans. 10:1

Q84. C:N Ratio of legume

Ans. 20: 1 to 30:1

Q85. Fastest N fixng plant

Ans. Sesbania rostrata

Q86. First Sequenced plant

Ans. Arabidopasis thaliana (a weed)

Q87. Bt was discovered by

Ans. Shigetane Ishiwatari

Q88. Term DNA Fingerprinting was coined by

Ans. Prof A. Jeffery

Q89. First Chemical pesticide used in India was

Ans. DDT

Q90. Which committee recommended ASRB

Ans. Gajendra Gadkar Committee

Q91. Food for work was started in year

Ans. 1977

Q92. Crop Insurance Scheme (CIS) was started in year

Ans. 1985

Q93. Firt hybrid of Pigeonpea was released in year

Ans. 1991

Q94. India first magazine contributed to agriculture

Ans. kheti

Q95. I.A.R.I was established in Bhiar

Ans. 1905

Q96. I.A.R.I was established under the vice royalty of

Ans. Lord Curzon in 1905

Q97. Builiding of I.A.R.I was damaged due to Earthquake

Ans. 1934

Q98. Transfer of IARR to new Delhi

Ans. 1936

Q99. ICAR was established by the recommendation of.

Ans. Lord Linlithgo

Q100. First President of ICAR was

Ans. Sir Mohammad Habibullah

Q101. First secretary of ICAR was

Ans. S.A. Hydari

Q102. First D.G of ICAR

Ans. Dr. B.P Pal

Q103. First Indian Director of IARI

Ans. B. Vishwanthan

Q104. First Vice chairman of ICAR

Ans. Diwan Bhadur Vijayraj Acharya

Q105. Vision 2030 launched in the year

Ans. January 2011

Q106. First agriculture minister of independent India

Ans. Rafi Ahmed Kidwai

Q107. Green revolution occurred In

Ans. 1965-1966

Q108. India's First fully organic state is

Ans. Sikkim

Q109. Microbial type culture collection centre is situated at

Ans. Chandigharh

Q110. How many type of element occurs in plant body

Ans. 20

Q111. Mycorrhize is the association of

Ans. Higher plants and fungi

Q112. The Study of soil fertility with respect to crop known as

Ans. Edopolgy

Q113. Element most mobile in soil?

Ans. Nitrogen

Q114. India's first biotech crop technology approved for commercialization in which year

Ans. 2002

Q115. Kisan credit card scheme started in which year

Ans. 1998

Q116. National genebank established at

Ans. New Delhi

Q117. NAAS (National Academy of Agri. Science) established

Ans. 1990

Q118. ICAR foundation day

Ans. 16 july

Q119. Phone number of Kisaan call centre

Ans. 18001801551

Q120. Most salt tolerant fruit crop is

Ans. Date palm

Q121. First tractor bought in India I which year

Ans. 1914

Q122. Non agriculture institute which has developed 40 varieties of crops

Ans. BARC

Q123. Age of plants is determind by

Ans. Annual rings

Q124. Agmark is

Ans. Quality of food product

Q125. The first fertilizer produced I India was

Ans. S.S.P

Q126. Back bone of America is the crop

Ans. Maize

Q127. Queen of Cereals known as

Ans. Maize

Q128. Queen of spices

Ans. Cardamom

Q129. First man made cereals is

Ans. Triticale

Q130. International Pest is

Ans. Schistacerca gregaria

Q131. Instrument used to Measure atmospheric pressure

Ans. Barometer

Q132. Who discovered Vitamin

Ans. Funk

Q133. National Farmers day is celebrated on

Ans. first Friday of dec

Q134. Plant known as a dairy plant also is

Ans. Soyabean

Q135. ICAR Normon Borlaug Award given in Time of Year

Ans. Once in 5 year

Q136. Property of soil Which can't be changed

Ans. Soil texture

Q137. The effective new fungicide for control of oomycetous fungus is

Ans. Metalaxyl

Q138. Plant have no economics importance for human being is

Ans. Absolute weed

Q139. C:N ratio for normal soil found are

Ans. C:N 10:1

Q140. Study of soil in relation with higher plants is called

Ans. Edaphology

Q141. Major disease of Sugarcane is

Ans. Red rot

Q142. State known as bowl of rice

Ans. Chhattisgarh

Q143. The gas emitted from rice

Ans. Methane

Q144. Most critical stage of water

Ans. Booting stage

Q145. SRI stand sfor

Ans. System of Rice Intensification.

Q146. SRI concept

Ans. Increasing productivity by changing management of plants, soil, water and nutrients.

Q147. Yield increased by adopting by SRI

Ans. 50-90%

Q148. ISOPOM scheme is launched in

Ans. 2004 in 14 major state

Q149. White revolution is related to

Ans. Milk

Q150. Blue revolution is related to

Ans. Fisheries

Q151. Round revolution is related to

Ans. Potato

Q152. Red revolution is related

Ans. Tomato

Q153. Pink revolution is related to

Ans. Onions

Q154. Rainbow revolution is related to

Ans. Overall development of agriculture

Q155. Golden revolution is related to

Ans. Horticulture (Dr. K.L. Chadha)

Q156. Yellow revolution is realted to

Ans. Oilseeds

Q157. Silver revolution is related to

Ans. Egg and poultry

Father of various Fields

Father of Agricultural chemistry-	Justus Von Liebig
Father of biology-	Aristole
Father of eugenics –	F. Galtons
Father of zoology	Aristole
Father of modern cell biology-	George Pallade
Father of genetics –	G.J mendel
Father of of modern genetics –	Bateson
Father of cytoplasmic inheritance –	Carl correns
Father of plant physiology-	Stephhen Hales
Father of Bacteriology-	Robert Koch
Father of fermentation –	Louis Pasteur
Father of taxonomy –	C. Linnaeus
Father of microscopic Anatomy-	Malphighi
Father of plant anatomy-	N. grew
Father of population genetics –	Sir R.A Fisher
Father of modern plant pathology –	Anton De bary
Father of mycology –	Micheli
Father of Virology –	M.W Beijernick
Father of plant tissue culture –	G. haberlandt
Father of genetics engineering –	Paul berg
Father of immunology –	Edwrd jenner
Father of vital statistics –	Captain John Grant
Father of sociaology –	Auguste Comte
Father of Agronomy –	Peter Decresenzi
Father of nematology –	Nathan Augustus Cobb.
Father of of green revolution -	Norman E. Borlaug
Father of Microbilogy-	Antony von leeuonweok
Father of Medical microbiology –	Robert koch
Father of DNA finger printing –	Alec jeffreys
Father of tissue culture –	Hrrisons
Father of epidemiology –	John Snow

Annexures

Annexure- I

Table: Diseases of crop plants and their causative agents

S. No.	Name of the Crop	Diseases	Causal agents
1.	Rice	Rice blast	*Pyricularia oryzae*
		Brown spot	*Helminthosporium oryzae/ Bipolaris oryzae*
		Sheath blight	*Rhizoctonia solani*
		Stem rot	*Sclerotium oryzae*
		False smut	*Ustilaginoidea virens*
		Bacterial leaf blight	*Xanthomonas oryzae*
		Bacterial leaf streak	*Xanthomonas oryzicola*
		Rice tungru	Virus
		Grassy stunt	Virus
		Rice yellow dwarf	Phytoplasma
	Wheat	Black/stem rust	*Puccinia graminis tritici*
		Yellow/stripe rust	*Puccinia striformis*
		Brown/orange rust	*Puccinia recondita*
		Loose smut	*Ustilago tritici*
		Bunt disease	*Tilletia carries*/T. foetida
		Karnal bunt	*Tilletia indica*
		Powdery mildew	*Erysiphe graminis* f. sp. *tritici*
		Leaf blight	*Alternaria triticina*
	Sorghum	Downy mildew	*Peronospora sorghi*
		Grain smut	*Sphacelotheca sorghi*
		Loose smut	*Sphacelotheca cruenta*
		Head smut	*Sphacelotheca reiliana*

	Long smut	*Tolysporium ehrenbergii*
Pearl millet/Bajra		Green ear/ downy mildew *Sclerospora graminicola*
	Ergot disease	*Claviceps fusiformis*
	Smut	*Tolyposporium penicillariae*
	Rust	*Puccinia penniseti*
Maize	Brown spot	*Zea maydis*
	Corn smut	*Ustilago maydis*
	Downy mildew	*Sclerospora sorghi*

Annexures-II

Table: Various Types of human Infections and their causative agents

Viral disease		
Airborne Diseases		
S. No.	**Disease**	**Causative agent**
1.	Chickenpox (varicella)	Herpesviridae virus
2.	Influenza (Flu)	Orthomyxoviruses
3.	Measles (Rubeola)	Virus belongs to the genus Morbillivirus and the family Paramyxoviridae.
4.	Mumps	Member of the genusRubulavirus in the family Paramyxoviridae.
5.	Respiratory Syndromes and Viral Pneumonia	The adenoviruses, Coxsackievirus A, Coxsackievirus B, echovirus,influenza viruses, parainfluenza viruses, poliovirus, respiratory syncytialvirus, and reovirus are thought to be responsible
6.	Rubella (German Measles)	Rubella virus, a member of the family Togaviridae
7.	Smallpox (Variola)	The variola virus belongs to the family Poxviridae
8.	Colorado Tick Fever	Virus belongs to the genus Coltivirus
9.	Yellow Fever	Flavivirus
10.	Acquired Immune Deficiency Syndrome (AIDS) Acquired Immune Deficiency Syndrome (AIDS)	Retroviridae

11.	Cold Sores	Herpessimplex virus type 1 (HSV-1)
12.	Common Cold	Picornaviridae
13.	Cytomegalovirus Inclusion Disease	Human cytomegalovirus(HCMV), a member of the family Herpesviridae
14.	Genital Herpes	Genital herpes is caused by the herpes simplex virus type 2(HSV-2). of the family Herpesviridae
15.	Human Herpesvirus 6 Infections	Human herpesvirus 6 (HHV-6)
16.	Human Parvovirus B19 Infections	Parvovirus
17.	Leukemia	By two retroviruses: humanT-cell lymphotropic virus I (HTLV-I) and HTLV-II. HTLV-Iand HTLV-II are members of the family Retroviridae.
18.	Mononucleosis (Infectious)	The Epstein-Barr virus (EBV)
19.	Rabies	Family Rhabdoviridae
20.	Viral Hepatitides	Two are herpesviruses (cytomegalo-virus [CMV] and Epstein-Barr virus [EBV]) and seven are hepatotropic viruses.
21.	Hepatitis B (serum hepatitis)	HBV is classified as an Ortho-hepadnavirus within thefamily Hepadnaviridae
22.	Hepatitis A (Hepatitis A)	Picornaviridae, Hepatovirus
23.	Hepatitis C (Hepatitis C)	Flaviviridae, Pestivirus
24.	Hepatitis D (Hepatitis D)	Unclassified
25.	Hepatitis E (Hepatitis E)	Caliciviridae (?)
26.		
27.	Poliomyelitis	Caused by the poliovirus, amember of the family Picornaviridae
Slow Virus and Prion Diseases of Humans		
28.	Creutzfeldt-Jakob disease (CJD)	Prion
29.	Kuru	Prion
30.	Gerstmann-Sträussler- Scheinker Syndrome (GSS)	Prion

31.	Fatal familial insomnia (FFI)	Prion
32.	Progressive multifocal leukoencephalopathy	Papovavirus
33.	Subacute sclerosing panencephalitis (SSPE)	Measle virus variant
	Other viral diseases	
34.	Warts or verrucae	Caused by the human papillomaviruses
	Bacterial Diseases	
	Airborne	
35.	Diphtheria	*Corynebacterium diphtheriae*
36.	Legionnaires' Disease and Pontiac Fever	*Legionella pneumophila*
37.	Meningitis	Streptococcus pneumoniaeNeisseria meningitidisHaemophilus influenzae type bGram-negative bacilli Group B streptococciListeria monocytogenes Mycobacterium tuberculosisNocardia asteroides Staphylococcus aureus Staphylococcus epidermidis
38.	Pertussis	*Bordetella pertussis*
39.	Cellulitis	*S.Pyogenes & S. aureus*
40.	Erysipelas	*S. Pyogenes*
41.	Scarlet Fever	*S. pyogenes*
42.	Streptococcal Pneumonia	*Streptococcus pneumoniae*
43.	Tuberculosis	*Mycobacterium tuberculosis*
	Arthropod-Borne Diseases	
44.	Ehrlichiosis	*Ehrlichia chaffeensis*
45.	Epidemic (Louse-Borne) Typhus	*Rickettsia prowazekii*
46.	Endemic (Murine) Typhus	*Rickettsia Rickettsia typhi*
47.	Lyme Disease	*Borrelia burgdorferi, B. garinii and B. afzelii.*
48.	Plague	*Yersinia pestis*
49.	Q Fever	*Coxiella burnetii*
50.	Rocky Mountain Spotted Fever	*Rickettsia rickettsii*

Direct Contact Diseases	
51. Anthrax	*Bacillus anthracis*
52. Bacterial Vaginosis	*Gardnerella vaginalis*
53. Cat-Scratch Disease	*Bartonella henselae*
54. Chancroid	*Haemophilus ducreyi*
55. Chlamydial Pneumonia	*Chlamydia pneumoniae*
56. Gas Gangrene or Clostridial Myonecrosis	*Clostridium perfringens*, *C. novyi* and *C. septicum*
57. Genitourinary Mycoplasmal Diseases	*Ureaplasma urealyticum* and *M. hominis*
58. Gonorrhea	*Neisseria gonorrhoeae*
59. Inclusion Conjunctivitis	*C. trachomatis*
60. Leprosy	*Mycobacterium leprae*
61. Lymphogranuloma Venereum	*Chlamydia trachomatis*
62. Mycoplasmal Pneumonia	*C. trachomatis, Ureaplasma urealyticum, Mycoplasma hominis, Trichomonas vaginalis, Candida albicans*
63. Nongonococcal Urethritis	*Neisseria gonorrhoeae*
64. Peptic Ulcer Disease and Gastritis	*Campylobacter pylori*
65. Psittacosis (Ornithosis)	*Chlamydia psittaci*
66. Syphilis	*Treponema pallidum*
67. Tetanus	*Clostridium tetani*
68. Trachoma	*Chlamydia trachomatis*
69. Tularemia	*Francisella tularensis*

Food-Borne and Waterborne Diseases	
70. Botulism	*Clostridium botulinum*
71. Campylobacter jejuni Gastroenteritis	*C. jejuni*
72. Cholera	*Vibrio cholerae*
73. Listeriosis	*Listeria monocytogenes*
74. Salmonellosis	*Salmonella serovars*
75. Shigellosis	*S. sonnei, S. flexneri*
76. Staphylococcal Food Poisoning	*Staphylococcus aureus*
77. Traveler's Diarrhea	*Escherichia coli*
78. Typhoid Fever	*Salmonella typhi*

Dental infections		
79.	Dental plaque	*Streptococcus gordonii, S. oralis*, and *S. mitis*
80.	Periodontal Disease	*Porphyromonasgingivalist*
Fungal diseases		
81.	Black piedra	*Piedraia hortae*
82.	White piedra	*Trichosporon beigelii*
83.	Tinea versicolor	*Malassezia furfur*
84.	Tinea barbae	*Trichophyton mentagrophytes, Tinea barbaeT. verrucosum, T. rubrum*
85.	Tinea capitis	*Trichophyton, Microsporum canis*
86.	Tinea corporis	*Trichophyton rubrum, T. mentagrophytes, Microsporum canis*
87.	Tinea cruris (jock itch)	*Epidermophyton floccosum, T. mentagrophytes, T. rubrum*
88.	Tinea pedis (athlete's foot)	*T. rubrum, T. mentagrophytes, E. floccosum*
89.	Tinea unguium (onychomycosis)	*T. rubrum, T. mentagrophytes, E. floccosum*
90.	Chromoblastomycosis	*Phialophora verrucosa, Fonsecaea pedrosoi*
91.	Maduromycosis	*Madurella mycetomatis*
92.	Sporotrichosis	*Sporothrix schenckii*
93.	Blastomycosis	*Blastomyces dermatitidis*
94.	Coccidioidomycosis	*Coccidioides immitis*
95.	Cryptococcosis	*Cryptococcus neoformans*
96.	Histoplasmosis	*Histoplasma capsulatum*
97.	Aspergillosis	*Aspergillus fumigatus, A. flavus*
98.	Candidiasis	*Candida albicans*
99.	Pneumocystis pneumonia	*Pneumocystis carinii*
Protozoan Diseases		
100.	Amebiasis	*E. histolytica*
101.	Amebic meningoencephalitis	*Acanthamoeba* spp., *Naegleria fowleri*
102.	Cryptosporidiosis	*Cryptosporidium parvum*
103.	Cyclosporidiosis	*Cyclospora cayetanensis*

104. Isosporiasis	*Isospora belli*
105. Toxoplasmosis	*Toxoplasma gondii*
106. Balantidiasis	*Balantidium coli*
107. Cutaneous leishmaniasis	*Leishmania tropica*
108. Mucocutaneous leishmaniasis	*L. braziliensis*
109. Kala-azar (visceral leishmaniasis)	*L. donovani*
110. American trypanosomiasis	*Trypanosoma cruzi*
111. African sleeping sickness	*T. brucei gambiense, T. brucei rhodesiense*
112. Giardiasis	*Giardia lamblia*
113. Trichomoniasis	*Trichomonas vaginalis*
114. Malaria	*Plasmodium falciparum, P. malariae, P. ovale, P. vivax*
115. Microsporidiosis	*Encephalitozoon, Nosema, Vittaforma, Pleistophora, Enterocytozoon, Trachipleistophora, Microsporidium*
116. Trypanosomiasis	*Trypanosoma brucei gambiense*
Major Sexually Transmitted Diseases (STDs) Viral disease	
117. Acquired immune deficiency syndrome(AIDS)	Human Immunodeficiency Virus (HIV)
118. Genital herpes	Herpes Simplex Virus (HSV-2)
119. Condyloma acuminata (genital warts)	Condyloma Acuminata (genital warts) (HPV-6)
120. Hepatitis B (serum hepatitis)	Hepatitis B virus (HBV)
121. Congenital cytomegalic inclusion disease	Cytomegalovirus (CMV)
122. Genital molluscum contagiosum	*Molluscum contagiosum*
Bacterial diseases	
123. Granuloma inguinale (donovanosis)	*Calymmatobacterium granulomatis*
124. Diarrhea and rectal inflammation inhomosexual men	*Campylobacter* (Heliobacter) *cinaedi, C. fennelliae*
125. Nongonococcal urethritis (NGU);cervicitis, pelvic inflammatorydisease (PID), lymphogranulomavenereum	*Chlamydia trachomatis*

126. Bacterial vaginosis	*Gardnerella vaginalis*
127. Chancroid ("soft chancre")	*Haemophilus ducreyi*
128. Implicated in some cases of NGU	*Mycoplasma genitalium*
129. Implicated in some cases of PID	*Mycoplasma hominis*
130. Gonorrhea, PID	*Neisseria gonorrhoeae*
131. Syphilis, congenital syphilis	*Treponema pallidum subsp. pallidum*
132. Urethritis	*Ureaplasma urealyticum*
Disease caused by Yeasts	
133. Candidiasis (moniliasis)	*Candida albicans*
Protozoan diseases	
134. Trichomoniasis	*Trichomonas vaginalis*